AF333501

Functional Polymer and Composite Systems

Volume 2

Functional Polymer and Composite Systems

Volume 2

Dr. Vikas Mittal

Editor and Lead Author

CWP

Central West Publishing

A catalogue record for this book is available from the National Library of Australia

ISBN (print): 978-1-922617-20-0

Contents

Preface   IX

**1.   Polymer Nanocomposites for Aerospace   1
        Applications**

        1.1   Introduction   1
        1.2   Development of Polymer Nanocomposites for   1
              Aerospace Applications
        1.3   Conclusions   8
        References   9

**2.   PNIPAAM Grafted Porous Polymeric Monoliths: 19
        Parameters Affecting Structure and Morphology**

        2.1   Introduction   19
        2.2   Experimental   21
                    2.2.1   Materials   21
                    2.2.2   Synthesis of Polystyrene Particles   21
                            and Surface Functionalization
                    2.2.3   Shear Aggregation   21
                    2.2.4   Monolith Synthesis   22
                    2.2.5   Electron Microscopy   22
        2.3   Results and Discussion   23
        2.4   Conclusions   29
        References   31

**3.   Polymer Nanocomposites for Automotive   33
        Applications**

        3.1   Introduction   33
        3.2   Polymer Nanocomposites for Automotive   34
              Tires
                    3.2.1   Styrene Butadiene Rubber (SBR)   35
                            Nanocomposites
                    3.2.2   Natural Rubber Nanocomposites   38
                    3.2.3   Butyl Rubber Nanocomposites   43
                    3.2.4   Other Elastomer Nanocomposites   44
        3.3   Polymer Nanocomposites for Automotive Tire   46
              Inner Tubes

3.4   Polymer Nanocomposites for Automotive   47
        Coatings
        3.4.1   Acrylic/Melamine Nanocomposite   48
                 Coatings
        3.4.2   Polyurethane Nanocomposite   48
                 Coatings
        3.4.3   Acrylic Nanocomposite Coatings   50
        3.4.4   Polyester Nanocomposite Coatings   51
3.5   Polymer Nanocomposites for Automotive   52
        Polymer Electrolyte Membrane Fuel Cells
3.6   Polypropylene (PP) Nanocomposites for   56
        Automotive Applications
3.7   Poly(lactic acid) (PLA) Nanocomposites for   56
        Automotive Applications
3.8   Conclusions   58
References   58

4.   Living Polymer Architectures   73

4.1   Introduction   73
4.2   Living/Controlled Polymerization Techniques   73
        4.2.1   Nitroxide Mediated Polymerization   73
        4.2.2   Living Anionic and Telluride   75
                 Mediated Polymerizations
        4.2.3   Reversible Addition Fragmentation   76
                 Chain Transfer
        4.2.4   Atom Transfer Radical   77
                 Polymerization
4.3   Advantages of Living Polymerization   79
        Techniques
4.4   Applications of Living Polymer Architectures   79
4.5   Conclusion   81
References   81

**5.   Epoxy: Understanding Degradation and   91
        Stabilization**

5.1   Introduction   91
5.2   Understanding the Degradation and   92
        Stabilization of Epoxy
        5.2.1   Thermal Degradation   92

5.2.2 Fire Retardation 95
5.2.3 UV Degradation 96
5.2.4 Epoxy Stabilization 96
5.3 Conclusions 99
References 99

6. Polymer Nanocomposites for Electronics Applications 105

6.1 Introduction 105
6.2 Electronics Applications of Various Polymer Nanocomposites 106
6.2.1 Graphene Based Polymer Nanocomposites 106
6.2.2 CNTs Based Polymer Nanocomposites 113
6.2.3 CNFs Based Polymer Nanocomposites 123
6.2.4 ZnO Based Polymer Nanocomposites 126
6.3 Summary 128
References 129

7. Polymer Nanocomposites for Sensing Applications 139

7.1 Introduction 139
7.2 Conducting Polymer Nanocomposites for Sensing 140
7.3 Development of Different Sensors 141
7.3.1 Gas Sensors 141
7.3.2 Biosensors 145
7.3.3 Humidity Sensors 149
7.3.4 Liquid Sensing 152
7.4 Conclusions 153
References 154

8. Liquid Crystalline Polymer Composites 167

8.1 Introduction 167
8.2 Liquid Crystals 167
8.3 Classification of LCPs 168

8.3.1   Main Chain LCPs   168
8.3.2   Side Chain LCPs   168
8.4   Liquid Crystalline Polymer Nanocomposites   169
8.5   Conclusion   172
References   173

9.   Polymer Nanocomposite Inks and Pigments   185

9.1   Introduction   185
9.2   Printing Techniques   186
9.3   Polymer Nanocomposites based Inks   186
9.4   Polymer Nanocomposites based Pigments   188
9.5   Conclusions   189
References   189

10.   Biopolymer Coatings   197

10.1   Introduction   197
10.2   Cellulose   198
10.3   Chitosan   202
10.4   Starch   212
10.5   Polylactide   213
10.6   Vegetable Oil Based Coatings   215
References   220

Index   231

Preface

A wide spectrum of polymer-based materials with optimal property profiles and performance have resulted owing to the recent advancements in polymer science and technology. The current book presents various aspects related to the development and characterization of a range of such polymeric systems and polymer composites with special emphasis on their applications.

In Chapter 1, the development and application of the functional polymer nanocomposites for aerospace applications have been reviewed in brief. Chapter 2 reports the synthesis of the monoliths by grafting of poly(N-isopropylacrylamide) (PNIPAAM) brushes from the aggregated particles or free latex particles and simultaneously crosslinking them in order to provide the monolith structure. Chapter 3 aims to summarize the recent developments with respect to the automotive applications of the polymer nanocomposites. In Chapter 4, different living polymerization techniques to achieve controlled polymer architectures have been briefly reviewed. In Chapter 5, the degradation behavior of epoxy is briefly studied by focusing on the mechanism of induced damage. Stabilization of epoxy with additives to overcome the degradation phenomena is also briefly introduced. In Chapter 6, recent advancements and prospects of polymer nanocomposites for use in the electronics industry are reviewed. Chapter 7 summarizes the recent advancements in the development of the polymer composite systems for use in various sensing applications. Chapter 8 reviews the synthesis and properties of the liquid crystalline polymer nanocomposite systems developed in the recent past. In Chapter 9, the development of polymer nanocomposite inks and pigments has been briefly reviewed. Chapter 10 presents an overview on the development of various biopolymer coatings and their ability in protecting metal substrates in different corrosive environments.

The book is a valuable reference for polymer, chemical and materials engineers as well as practicing scientists and researchers in both academia and industry. It is also a useful resource for postgraduate university students in chemical, polymer and materials disciplines, along with interdisciplinary streams.

The book would not have been successfully accomplished without the support of chapter contributors. The book is dedicated to my family for unswerving support, constant motivation and constructive suggestions for improvement.

Vikas MITTAL

1

Polymer Nanocomposites for Aerospace Applications

1.1 Introduction

Over the past decades, significant research efforts have been devoted for achieving optimal property enhancements in polymer nanocomposites. As compared to pure polymer matrices and micro-composites, nanocomposites exhibit extensive improvements in performance at very low amount of filler fractions [1]. Improved modulus and strength, enhanced gas barrier behavior along with superior adaptive and thermal properties are few features observed for nanocomposites [2-12]. Especially for aerospace applications, the major challenge includes the achievement of composites with multi-functionality having superior mechanical behavior and efficient flame retardation simultaneously [13-15].

An array of nanoscale fillers having diverse dimensional features is being used for the synthesis of polymer nanocomposites. With the surface to volume ratio of additives, it is possible to characterize the efficiency of the reinforcement [16]. Composites based on functional nanofillers have been observed to provide various multi-functional characteristics like thermal and dimensional stability, fire retardancy, field emission, electronic and optical properties, material durability, impact resistance, energy absorption, among others, which are particularly significant for a variety of aerospace applications [17-22].

In this chapter, the development and application of the functional polymer nanocomposites for aerospace applications has been reviewed in brief.

1.2 Development of Polymer Nanocomposites for Aerospace Applications

As compared to the commercial carbon fibers with a compressive strain to failure of almost 1%, carbon fiber reinforced polymer composites, utilized for aerospace applications, are usually allocated an

Swati Singh and Vikas Mittal, Khalifa University of Science and Technology, Abu Dhabi, UAE

acceptable compressive strain level of <0.4% [23,24]. The accurate detection of damage and cracks in the composite materials are, thus, needed to avoid safety hazards as well as to reduce maintenance costs. A number of methodologies have also been developed to achieve self-healing of polymer materials [25-29]. These methods include microcapsules or glass tubes, having a healing agent that is unconstrained in an impairment location upon fracture. A certain extent of mechanical performance is then restored by the polymerization of the healing agent. A number of research studies have also been performed for the application of these systems in fiber-reinforced polymers [30-37]. For instance, Hucker *et al.* [38-41] have reported the use of resin filled hollow glass fibers (HGF) with diameters in the range of 30-100 μm and almost 50% hollowness for achieving self-healing. In another such study, Williams *et al.* [42] reported carbon fiber-reinforced epoxy with resin filled embedded HGF with self-healing functionality. Recovery was observed in flexural strength in laminates with resin filled HGF at two different fiber spacings (70 μm and 200 μm). With hollow glass fibers at 70 μm spacing, an optimum flexural strength recovery of 89% of the baseline laminate performance and 97% of the unharmed state performance was accomplished.

As mentioned earlier, to meet the application needs, the flame retardancy of polymer resins also needs to be enhanced [43-48]. A significant improvement in the flame retardancy of the epoxy system incorporated with carbon fibers was noted by Toldy *et al.* [49]. In order to achieve optimal enhancement in flammability and mechanical properties simultaneously, the authors fabricated multi-layer composites comprising of reference intumescent epoxy resin coating layer and composite core.

Liu *et al.* [50] also reported mixing of a tetra functional epoxy (TFTE) with diglycidyl ether of bisphenol A (DGEBA) to develop high performance composite material for aerospace applications. At 30 wt% TFTE content, glass transition temperature (T_g), impact strength and tensile strength were observed to be 252 °C, 29.8 kJ/m^2 and 80 MPa, respectively. In addition, the cured hybrid exhibited good tensile strength and interfacial shear strength with carbon fiber incorporation. Figure 1.1 shows the tensile and impact strength of the materials as a function of TFTE fraction. In another study, Rice *et al.* [51] prepared exfoliated epoxy nanocomposites with carbon fibers and organo-clays. The nano-clay was observed to be oriented parallel to the fiber axis, thus, resulting in optimal performance.

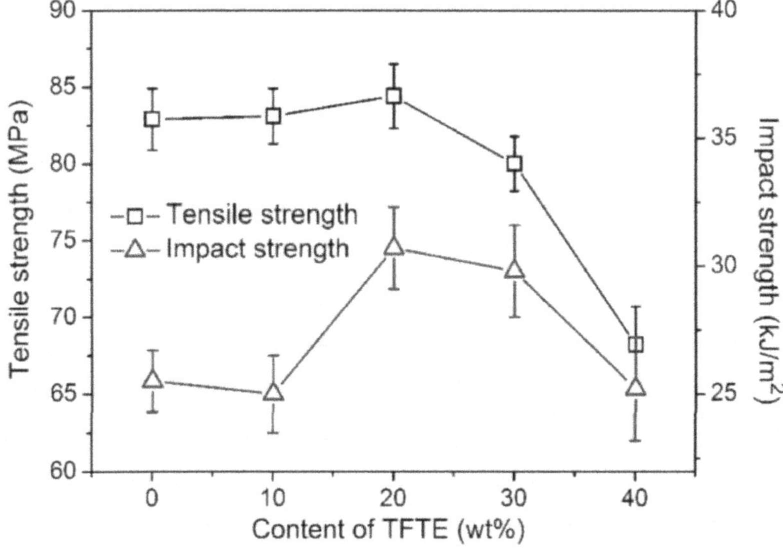

Figure 1.1 Tensile and impact strength as a function of TFTE content. Reproduced from Reference 50 with permission from American Chemical Society.

E-glass fiber-epoxy composites seem to be the first category of composite materials developed with healing capabilities, which followed the development of raw self-healing epoxy resins [52,53]. E-glass fiber-epoxy composites are extensively used due to their lightweight, affordable cost and superior mechanical properties. Numerous applications of E-glass fiber-epoxy composites include boats, sport cars bodies and aircraft components. The self-healing nature of E-glass fiber-epoxy composites with a dissolved poly(bisphenol A-co-epichlorohydrin) thermoplastic has been investigated [54]. The healing phenomenon was intrinsic, as the thermoplastic polymer was efficiently dispersed in the epoxy matrix. A recovery of up to 70% of the virgin material properties was observed, and the observed delamination area was also reduced. In another study, Yin *et al.* [55] reported self-healing woven glass fabric-reinforced epoxy composite laminates, using two-component healing system comprising of urea formaldehyde microcapsules containing epoxy and $CuBr_2(2\text{-MeIm})_4$ hardener. A healing efficiency of 68% was received. In another recent study, Zainuddin *et al.* [56] reported recovery and improvement of low-velocity impact properties of self-healing E-glass-epoxy composites generated by embedding HGF filled with self-healing agent

(SHA). A gain of 53.6% in peak load after second impact was observed in SHA filled HGFs samples. In addition, a gain of 86.6% in energy to peak load was also observed. Das *et al.* [57] have also reviewed different self-healing methodologies, self-healing polymers and ceramic and metal composites, useful for application in advanced aerospace structures.

Kong [58] reported volume recovery effects in carbon fiber-reinforced epoxy composites. For times up to 10^5 minutes, specimens after curing of Fiberite 934 epoxy resin as well as Thornel 300 carbon-fiber/Fiberite 934 epoxy were subjected to quenching from above glass transition temperature and annealed at 140 °C, 110 °C or 80 °C. No apparent weight loss was observed for the specimens during annealing. In another study, Kessler *et al.* [59] demonstrated self-healing for a structural carbon fiber-reinforced epoxy matrix composite. In this study, autonomic self-healing was achieved by adding dicyclopentadiene healing agent embedded in urea-formaldehyde microcapsules by *in-situ* polymerization. The composite was observed to recover up to 80% of its virgin interlaminar fracture toughness after delamination. Norris *et al.* [60] investigated embedded bioinspired vascular systems in carbon fiber-reinforced composites. The composite was developed as a aerospace-grade unidirectional carbon fiber-reinforced epoxy pre-impregnated tape with a [-45/90/45/0]2S sequence. The vascular network was implemented between plies using solder wire preforms. The healing was observed to be efficient when the vascules were located between the plies.

Wang *et al.* [61] studied a carbon fiber-reinforced epoxy matrix composite with poly(ethylene-co-methyl acrylate) (EMA) and poly(ethylene-co-methacrylic acid) (EMAA) patches placed between the plies. An improvement in the interlaminar fracture toughness, but a decrease in interlaminar shear strength were observed. Additionally, EMAA was observed to be a good choice for thermoplastic interlayer patches. In another study, Hargou *et al.* [62] reported a novel ultrasonic welding method to activate EMAA in composites. Partial healing of delamination cracks was observed, which resulted in complete recovery of delamination toughness.

The fiber-matrix interface in a carbon fiber-furan-functionalized epoxy matrix composite was also investigated by Zhang *et al.* [63]. Maleimide groups were also grafted on carbon fibers. Healing was achieved by a thermally activated Diels-Alder reaction between maleimide and furan groups. An increase in interfacial shear strength in comparison with untreated carbon fiber-epoxy system was observed.

The healing efficiency of the composites was observed to be 82% higher than the reference untreated composite. Other studies have also reported the use of electrical resistive heating to activate the healing process [57,64,65].

Polyetherimide (PEI), due to its low coefficient of thermal expansion and superior mechanical, thermal and electrical properties, is widely employed for specialty applications in electronics, aerospace and automotive industries. PEI is a homogeneous thermoplastic polymer with a glass transition temperature of 217 °C and has a thermal stability from cryogenic temperatures up to 350 °C. Due to its fluidity characteristics, it can be processed by injection molding, injection blow molding, extrusion and foaming. It is characteristically a flame resistant polymeric material with low emission of smoke. The modulus remains exceptionally high even at elevated temperatures. The presence of high concentration of aromatic groups in the aromatic polyetherimide helps in enhancing the radiation stability of polymer. These aromatic groups have the proficiency to convert the absorbed irradiated energy to heat, thus, diminishing the degree of chemical bond scission reaction. Carbon nanotubes (such as multi-walled carbon nanotubes (MWCNTs)) are commonly employed to enhance the mechanical properties, fire retardancy, thermal conductivity, electromagnetic shielding and irradiation shielding effectiveness [66-81]. Pitchan *et al.* [81,82] studied the effect of COOH functionalized MWCNTs as reinforcement in polyetherimide nanocomposites prepared by melt compounding. The authors suggested that nanofiller reinforced PEI is a functional material suitable to decrease the payload in aerospace structures. A content of 2 wt% of MWCNTs in PEI was observed to improve the mechanical properties by 15%. Activation energy for thermal degradation was also observed to increase by 699 kJ/mol for 2 wt% MWCNTs composite, indicating enhancement in thermal properties. In another study, Park *et al.* [83] fabricated a nanocomposite based on polyimide reinforced with homogenously dispersed single walled nanotubes (SWCNTs) for spacecraft applications. The nanocomposite exhibited superior thermal emissivity, high optical transparency, low solar absorptivity, electrical conductivity suitable to alleviate the static charge build-up, high thermal stability, low color in thin films as well as enhanced strength and stiffness.

A two-step twin screw extrusion process was used for the production of MWNTs/poly(p-phenylene sulfide) (PPS) nanocomposites by Noll *et al.* [84]. Electrical percolation threshold was observed below 0.77 vol% (1 wt%) filler content. In another study, Diez-Pascual *et al.*

[85] reported SWCNTs buckypaper (BP) embedded in poly(ether ether ketone) (PEEK) or PPS, which resulted in the improvement of mechanical and electrical properties as well as thermal conductivity. Compared to the neat polymers, considerably higher storage modulus and T_g were observed for the composites. Strong SWCNs-matrix interfacial adhesion was attributed to the remarkable enhancements in strength and stiffness. Yu *et al.* [86] also reported CNTs reinforced PPS nanocomposites, where electrical percolation was observed between 1-2 wt% filler content. For PPS nanocomposite with 5 wt% of CNTs, the onset of degradation temperature was observed to increase by 13.5 °C, whereas the modulus enhanced by 33%. The tensile strength of the composite also exhibited a significant enhancement of 172%.

Diez-Pascual *et al.* [87] reported the tribological (Figure 1.2) and

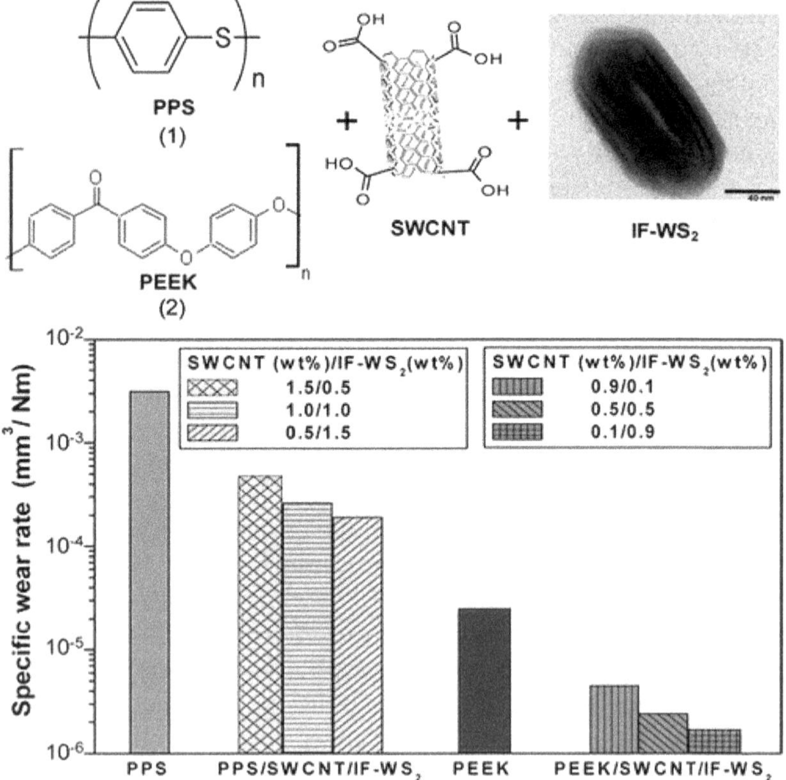

Figure 1.2 Schematic of composite synthesis and tribological properties of the composites. Reproduced from Reference 87 with permission from American Chemical Society.

rheological properties of SWCNTs-reinforced PPS and PEEK. The authors also evaluated the impact of dual-nanofiller strategy combining polyetherimide (PEI)-wrapped SWCNTs with inorganic fullerene-like tungsten disulfide (IF-WS$_2$) nanoparticles. The properties such as complex viscosity as well as storage and loss modulus were observed to enhance with SWCNTs content. The incorporation of increasing IF-WS$_2$ contents resulted in a corresponding decrease in viscosity and storage modulus, due to a lubricant effect. Both nanofillers resulted in improved wear behavior, which was attributed to the tribological properties of the nanoparticles and reinforcement effect.

Rana *et al.* [88-89] dispersed vapor-grown carbon nanofibers (CNFs) in carbon fabric-reinforced epoxy to develop three phase composites. High-speed mechanical stirring, ultrasonic treatment, among other manufacturing techniques, were analyzed for dispersing CNFs in epoxy. Incorporation of 0.5 wt% of CNFs was observed to significantly enhance mechanical, electrical and thermal properties of the three phase composites [90].

Another potential application of polymer nanocomposites for aerospace applications is aerostats. An aerostat is an advanced inflatable coated or laminated textile structure shaped like an aircraft, which lifts above the sea level at almost 3000 feet. The applications of such aerostats include raised stages for numerous electronic shipments extending from airborne early-warning radar systems to very low-frequency and low-frequency transportations, public emergency broadcasting equipment, communications relay, active or passive electronic warfare equipment, etc. Aerostats normally use helium gas to stay upward unlike helicopters or fixed-wing aircraft. The strength layer in a usual structure of inflatable aerostat is made of woven textiles, along with a protective layer which works as a gas barrier layer for retaining the inflated condition of the architecture for extended period of time. Thermoplastic polyurethane has gained significant attention as a material for the protecting layer of the aerostat [91]. However, in presence of destructive atmosphere (e.g. oxidative atmosphere, thermal exposure and UV radiation) for prolonged time, changes in mechanical, chemical and physical features of the polymer are observed [92]. UV stabilizers, antioxidants and UV absorbers are generally amalgamated into the formulation in order to evade dilapidation, however, such additives undergo occasional leaching, physical loss by migration, blooming, etc. Thus, the introduction of nanomaterials to generate functional polymer nanocomposites is a more functional approach to overcome the issue [93-95].

Bhattacharyya and Joshi [96] also reported Fe-Ni coating on acid functionalized nano-graphite through a fluidized bed electroplating procedure. Afterwards, the nanoparticles were distributed in polyurethane matrix and cast as nanocomposite films. An efficient microwave absorption was observed in the case of nano-graphite with thin Fe-Ni coating. The developed materials portray high potential for use in numerous defense applications like nets, camouflage cover, domain coatings for fighter airplanes and radomes, etc.

In another study, Guo *et al.* [97] reported self-healing of SiO_2 microcapsules based on cationic polymerization for potential aerospace coatings (epoxy based) applications (Figure 1.3). Epoxy resin and cationic photoinitiator were successfully encapsulated in the microcapsules via combined interfacial/*in-situ* polymerization.

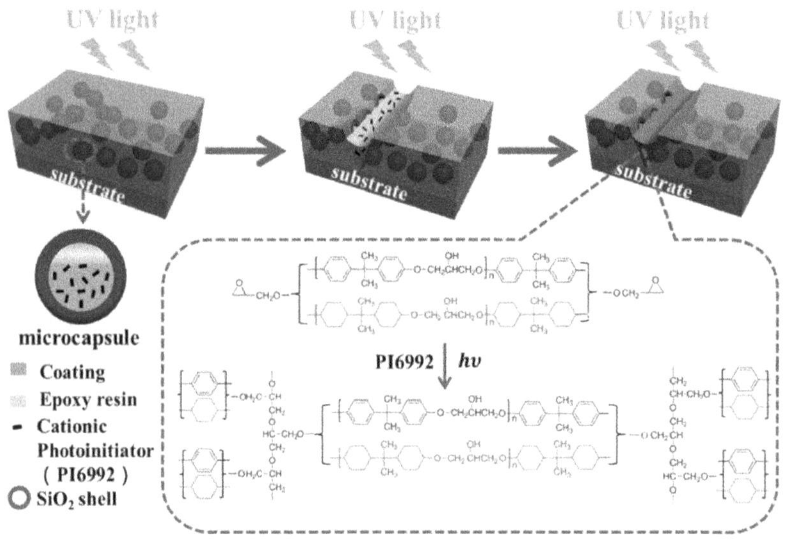

Figure 1.3 Schematic of UV-triggered self-healing of SiO_2 microcapsules embedded in epoxy resin. Reproduced from Reference 97 with permission from American Chemical Society.

1.3 Conclusions

Polymer nanocomposites provide considerably improved thermal performance, gas barrier, dynamic mechanical properties, resistance to small molecule permeation and ablative behavior, which make them very attractive materials for diverse aerospace applications. Incorporation of self-healing functionality further enhances the potential of such polymer nanocomposites. In this chapter, a brief overview on the development and potential of polymer nanocomposites for aerospace applications has been presented. It is envisaged that the ongoing research efforts in this direction would further enhance the fundamental know-how to tune the properties and performance of these materials, especially for aerospace applications.

References

1. Lincoln, D. M., Vaia, R. A., Beown, J. M., and Tolle, T. H. B. (2000) Revolutionary nanocomposite materials to enable space systems in the 21st century. *IEEE Aerospace Conference Proceedings*, **4**, 183-192.
2. Kim, B. K., Seo, J. W., and Jeong, H. M. (2003) Morphology and properties of waterborne polyurethane/clay nanocomposites. *European Polymer Journal*, **39**, 85-91.
3. Patton, R. D., Pittman, Jr., C. U., Wang, L., and Hill, J. R. (1999) Vapor grown carbon fiber composites with epoxy and poly(phenylene sulfide) matrices. *Composites, Part A: Applied Science and Manufacturing*, **30**, 1081-1091.
4. Smith, Jr., J. G., Delozier, D. M., Connell, J. W., and Watson, K. A. (2004) Carbon nanotube-conductive additive-space durable polymer nanocomposite films for electrostatic charge dissipation. *Polymer*, **45**(18), 6133-6142.
5. Mittal, V. (2009) Polymer layered silicate nanocomposites: A review. *Materials*, **2**, 992-1057.
6. Sandler, J., Werner, P., Shaffer, M. S. P., Demchuk, V., Altstadt, V., and Windle, A. H. (2002) Carbon-nanofibre-reinforced poly(ether ether ketone) composites. *Composites Part A: Applied Science and Manufacturing*, **33**, 1033-1039.
7. Andrews, R., Jacques, D., Qian, D., and Rantell, T. (2002) Multiwall carbon nanotubes: Synthesis and application. *Accounts of Chemical Research*, **35**, 1008-1017.
8. Njuguna, J., and Pielichowski, K. (2003) Polymer nanocomposites for aerospace applications: Fabrications, *Advanced Engineering Materials*, **6**, 193-203.
9. *Delaware Composites Design Encyclopedia*, Technomic Publishing Company, USA (1990).
10. Odegard, G. M., Gates, T. S., Nicholson, L. M., and Wise, K. E. (2001) Equivalent-Continuum Modeling With Application to Carbon Nanotubes. NASA, USA. Online: Equivalent-Continuum Modeling With Application to Carbon Nanotubes (assessed 19th September 2018).

11. Star, A., Stoddart, J. F., Steuerman, D., Diehl, M., Boukai, A., Wong, E. W., Yang, X., Chung, S.-W., Choi, H., and Heath, J. R. (2001) Preparation and properties of polymer-wrapped single-walled carbon nanotubes. *Angewandte Chemie, International Edition*, **40**, 1721-1725.

12. Odegard, G. M., Gates, T. S., Wise, K. E., Park, C., and Siochi, E. J. (2002) Constitutive modeling of nanotube–reinforced polymer composites. *Composites Science and Technology*, **63**(11), 1671-1687.

13. Lan, T., and Beyer, G. (2011) Introduction to flame retardancy of polymer-clay nanocomposites. In: *Thermally Stable and Flame Retardant Polymer Nanocomposites*, Mittal, V. (ed.), Cambridge University Press, UK, pp. 161-185.

14. Njuguna J., and Pielichowski K., (2003) Polymer nanocomposites for aerospace applications: Characterization. *Advanced Engineering Materials*, **6**, 204-210.

15. Njuguna, J., and Pielichowski, K. (2003) Polymer nanocomposites for aerospace applications: Properties. *Advanced Engineering Materials*, **5**, 769-778.

16. McCrum, N. G., Buckley, C. P., and Bucknall, C. B. (1996) *Principles of Polymer Engineering*, Oxford Science Publications, USA.

17. Thostenson, E. T., Li, C., and Chou, T.-W. (2005) Nanocomposites in context. *Composites Science & Technology*, **65**, 491-516.

18. Cui, C.-H., Yan, D.-X., Pang, H., Xu, X., Jia, L.-C., and Li, Z.-M., (2016) Formation of segregated electrically conductive network structure in low-melt-viscosity polymer for highly efficient electromagnetic interference shielding. *ACS Sustainable Chemistry and Engineering*, **4**, 4137-4145.

19. Cha, J., Jin, S., Shim, J. S., Park, C. S., Ryu, H. J., and Hong, S. H. (2016) Functionalization of carbon nanotubes for fabrication of CNT/epoxy nanocomposites. *Materials & Design*, **95**, 1-8.

20. Alexandre, M., and Dubois, P. (2000) Polymer-layered silicate nanocomposites: preparation, properties and uses of a new class of materials. *Materials Science and Engineering R: Reports*, **28**, 1-63.

21. Fang, F. F., Choi, H. J., and Joo J. (2008) Conducting polymer/clay nanocomposites and their applications. *Nanoscience and Nanotechnology*, **8**, 1559-1581.

22. Gangopadhyay, R., and De, A. (2000) Conducting polymer nanocomposites: A brief overview. *Chemistry of Materials*, **12**, 608-622.

23. Zhou, G. (1998) The use of experimentally-determined impact force as a damage measure in impact damage resistance and tolerance of composite structures. *Composite Structures*, **42**, 375-382.

24. Richardson, M. O. W., and Wisheart, M. J. (1996) Review of low-velocity impact properties of composite materials. *Composites, Part A: Applied Science and Manufacturing*, **27**, 1123-1131.

25. Dry, C. (1996) Procedures developed for self-repair of polymer matrix composites. *Composite Structures*, **35**, 263-269.
26. Chen, X., Dam, M. A., Ono, K., Mal, A., Shen, H., Nutt, S. R., Sheran, K., and Wudl, F. (2002) A thermally re-mendable cross-linked polymeric material. *Science*, **295**, 1698-1702.
27. Zako, M., and Takano, N. (1999) Intelligent materials systems using epoxy particles to repair microcracks and delamination in GFRP. *Journal of Intelligent Material Systems and Structures*, **10**, 836-841.
28. Yuan, Y. C., Yin, T., Rong, M. Z., and Zhang, M. Q. (2008) Self healing in polymers and polymer composites. Concepts, realization and outlook: A review. *Express Polymer Letters*, **2**(4), 238-250.
29. Brown, E. N., White, S. R., and Sottos, N. R. (2004) Microcapsule induced toughening in a self-healing polymer composite. *Journal of Materials Science*, **39**, 1703-1710.
30. Motoku, M., Vaidya, U. K., and Janowski, G. M. (1999) Parametric studies on self-repairing approaches for resin infused composites subjected to low velocity impact. *Smart Materials and Structures*, **8**, 623-638.
31. Bleay, S. M., Loader, C. B., Hawyes, V. J., Humberstone, L., and Curtis, P. T. (2001) A smart repair system for polymer matrix composites. *Composites, Part A: Applied Science and Manufacturing*, **32**, 1767-1776.
32. Kessler, M. R., and White, S. R. (2000) Self-activated healing of delamination damage in woven composites. *Composites, Part A: Applied Science and Manufacturing*, **32**, 683-699.
33. Scheiner, M., Dickens, T. J., and Okoli, O. (2016) Progress towards self-healing polymers for composite structural applications. *Polymer*, **83**, 260-282.
34. Brown, E. N., White, S. R., and Sottos, N. R. (2005) Retardation and repair of fatigue cracks in a microcapsule toughened epoxy composite - Part II: In situ self-healing. *Composites Science and Technology*, **65**, 2474-2480.
35. Pang, J. W. C., and Bond, I. P. (2005) 'Bleeding composites' - damage detection and self-repair using a biomimetic approach. *Composites, Part A: Applied Science and Manufacturing*, **36**, 183-188.
36. Pang, J. W. C., and Bond, I. P. (2005) A hollow fibre reinforced polymer composite encompassing self-healing and enhanced damage visibility. *Composites Science and Technology*, **65**, 1791-1799.
37. Trask, R. S., and Bond, I. P. (2006) Biomimetic self-healing of advanced composite structures using hollow glass fibres. *Smart Materials and Structures*, **15**, 704-710.
38. Hucker, M., Bond, I. P., Foreman, A., and Hudd, J. (1999) Optimisation of hollow glass fibres and their composites. *Advanced Composite Letters*, **8**, 181-189.
39. Hucker, M. J., Bond, I. P., Haq, S., Bleay, S., and Foreman, A. (2002)

Influence of manufacturing parameters on the tensile strengths of hollow and solid glass fibres. *Journal of Materials Science*, **37**, 309-315.

40. Hucker, M., Bond, I., Bleay, S., and Haq, S. (2003) Investigation into the behavior of hollow glass fibre bundles under compressive loading. *Composites, Part A: Applied Science and Manufacturing*, **34**, 1045-1052.

41. Hucker, M., Bond, I., Bleay, S., and Haq, S. (2003) Experimental evaluation of unidirectional hollow glass fibre/epoxy composites under compressive loading. *Composites, Part A: Applied Science and Manufacturing*, **34**, 927-932.

42. Williams, G., Trask, R., and Bond, I. (2007) A self-healing carbon fibre reinforced polymer for aerospace applications. *Composites, Part A: Applied Science and Manufacturing*, **38**, 1525-1532.

43. Kandola, B. K., Horrocks, A. R., Myler, P., and Blair, D. (2003) Mechanical performance of heat/fire damaged novel flame retardant glass-reinforced epoxy composites. *Composites, Part A: Applied Science and Manufacturing*, **34**, 863-873.

44. Mouritz, A. P., Feih, S., Kandare, E., Mathys, Z., Gibson, A. G., Des Jardin, P. E., Case, S. W., and Lattimer, B. Y. (2009) Review of fire structural modelling of polymer composites. *Composites, Part A: Applied Science and Manufacturing*, **40**, 1800-1814.

45. Xu, Y., and Van Hoa, S. (2008) Mechanical properties of carbon fiber reinforced epoxy/clay nanocomposites. *Composites Science and Technology*, **68**, 854-861.

46. Bekas, D. G., Tsirka, K., Baltzis, D., and Paipetis, A. S. (2016) Self-healing materials: A review of advances in materials, evaluation, characterization and monitoring techniques. *Composites, Part B: Engineering*, **87**, 92-119.

47. Braun, U., Balabanovich, A. I., Schartel, B., Knoll, U., Artner, J., Ciesielski, M., Doring, M., Perez, R., Sandler, J. K. W., Altstadt, V., Hoffmann, T., and Pospiech, D. (2006) Influence of the oxidation state of phosphorus on the decomposition and fire behaviour of flame-retarded epoxy resin composites. *Polymer*, **47**, 8495-8508.

48. Formicola, C., De Fenzo, A., Zarrelli, M., Frache, A., Giordano, M., and Camino, G. (2009) Synergistic effects of zinc borate and aluminium trihydroxide on flammability behaviour of aerospace epoxy system. *Express Polymer Letters*, **3**, 376-384.

49. Toldy, A., Szolnoki, B., and Marosi Gy. (2011) Flame retardancy of fibre-reinforced epoxy resin composites for aerospace applications. *Polymer Degradation and Stability*, **96**, 371-376.

50. Liu, T., Zhang, L., Chen, R., Wang, L., Han, B., Meng, Y., and Li, X., (2017) Nitrogen-free tetrafunctional epoxy and its DDS-cured high performance matrix for aerospace applications. *Industrial & Engineering Chemistry Research*, **56**(27), 7708-7719.

51. Rice B. P., Chen, C., Cloos, L., and Curliss, D. (2001) Carbon fiber composites: Organoclay-aerospace epoxy nanocomposites, Part I. *SAMPE Journal*, **37**(5), 7-9.
52. White, S. R., Sottos, N. R., Geubelle, P. H., Moore, J. S., Kessler, M. R., Sriram, S. R., Brown, E. N., and Viswanathan, S. (2001) Autonomic healing of polymer composites. *Nature*, **409**, 794-797.
53. Brown, E. N., Moore, J. S., White, S. R., and Sottos, N. R. (2003) Fracture and Fatigue Behavior of a Self-healing Polymer Composite. *Materials Research Society Symposium Proceedings, Volume 735*. Online: http://autonomic.beckman.illinois.edu/brown_files/Eric-BrownMRS2002.pdf (assessed 18th September 2018).
54. Hayes, S. A., Jones, F. R., Marshiya, K., and Zhang, W. (2007) A self-healing thermosetting composite material. *Composites, Part A: Applied Science and Manufacturing*, **38**, 1116-1120.
55. Yin, T., Rong, M. Z., Zhang, M. Q., and Yang, G. C. (2007) Self-healing epoxy composites: Preparation and effect of the healant consisting of microencapsulated epoxy and latent curing agent. *Composites Science and Technology*, **67**, 201-212.
56. Zainuddin, S., Arefin, T., Fahim, A., Hosur, M. V., Tyson, J. D., Kumar A., Trovillion, J., and Jeelani, S. (2014) Recovery and improvement in low-velocity impact properties of e-glass/epoxy composites through novel self-healing technique. *Composite Structures*, **108**, 277-286.
57. Das, R., Melchior, C., and Karumbaiah, K. M. (2016) Self-healing composites for aerospace applications. In: *Advanced Composite Materials for Aerospace Engineering*, Rana, S., and Fangueiro, R. (eds.), Woodhead Publishing, UK, pp. 334-364.
58. Kong, E. S. W. (1984) Volume recovery in aerospace epoxy resins: Effects on time-dependent properties of carbon fiber-reinforced epoxy composites. In: *Characterization of Highly Cross-linked Polymers*, Labana, S. S., and Dickie, R. A. (eds.), pp. 125-164.
59. Kessler, M. R., Sottos, N. R., and White S. R. (2003) Self-healing structural composite material. *Composites Part A: Applied Science and Manufacturing*, **34**, 743-753.
60. Norris, C. J., Bond, I. P., and Trask, R. S. (2011) The role of embedded bioinspired vasculature on damage formation in self-healing carbon fibre reinforced composites. *Composites, Part A: Applied Science and Manufacturing*, **42**, 639-648.
61. Wang, C. H., Sidhu, K., Yang, T., Zhang, J., and Shanks, R. (2012) Interlayer self-healing and toughening of carbon fibre/epoxy composites using copolymer films. *Composites, Part A: Applied Science and Manufacturing*, **43**, 512-518.
62. Hargou, K., Pingkarawat, K., Mouritz, A. P., and Wang, C. H. (2013) Ultrasonic activation of mendable polymer for self-healing carbon-epoxy laminates. *Composites, Part B: Engineering*, **45**, 1031-1039.

63. Zhang, W., Duchet, J., and Gerard, J. F. (2014) Self-healable interfaces based on thermo-reversible Diels-Alder reactions in carbon fiber reinforced composites. *Journal of Colloid and Interface Science*, **430**, 61-68.

64. Park, J. S., Darlington, T., Starr, A. F., Takahashi, K., Riendeau, J., and Hahn, H. T., (2010) Multiple healing effect of thermally activated self-healing composites based on Diels-Alder reaction. *Composites Science and Technology*, **70**, 2154-2159.

65. Heo, Y., and Sodano, H. A. (2015) Thermally responsive self-healing composites with continuous carbon fiber reinforcement. *Composites Science and Technology*, **118**, 244-250.

66. Zhang, T., Yang, J., Zhang, N., Huang, T., and Wang, Y. (2017) Achieving large dielectric property improvement in poly(ethylene-vinylacetate)/thermoplastic polyurethane/multiwall carbon nanotube nanocomposites by tailoring phase morphology. *Industrial & Engineering Chemistry Research*, **56**, 3607-3617.

67. Verge, P., Benali, S., Bonnaud, L., Minoia, A., Mainil, M., Lazzaroni, R., and Dubois, P. (2012) Unpredictable dispersion states of MWNTs in HDPE: a comparative and comprehensive study. *European Polymer Journal*, **48**, 677-683.

68. Xie, X. L., Mai, Y.-W., and Zhou, X.-P., (2005) Dispersion and alignment of carbon nanotubes in polymer matrix, a review. *Materials Science and Engineering R: Reports*, **49**, 89-112.

69. Dombovari, A., Halonen, N., Sapi, A., Szabo, M., Toth, G., Maklin, J., Kordas, K., Juuti J., Jantunen H., Kukovecz A., and Konya Z. (2010) Moderate anisotropy in the electrical conductivity of bulk MWCNT/epoxy composites. *Carbon*, **48**, 1918-1925.

70. Kukovecz, A., Kozma, G., and Konya, Z. (2013) Multiwalled carbon nanotubes. In: Springer Handbook of Nanomaterials, Vajtai, R. (ed.), Springer, Germany, pp. 147-188.

71. Liu, T., Tong, Y., and Zhang, W.-D., (2007) Preparation and characterization of carbon nanotube/polyetherimide nanocomposite films. *Composites Science and Technology*, **67**, 406-412.

72. Li, C.-C., Lin, J.-L., Huang, S.-J., Lee, J.-T., and Chen, C.-H. (2007) A new and acid-exclusive method for dispersing carbon multi-walled nanotubes in aqueous suspensions. *Colloids and Surfaces A: Physicochemical and Engineering Aspects*, **297**, 275-281.

73. Li, L., and Xing, Y. (2007) Pt-Ru nanoparticles supported on carbon nanotubes as methanol fuel cell catalysts. *Journal of Physical Chemistry C*, **111**(6), 2803-2808.

74. Raffa, V., Vittorio, O., Riggio, C., Ciofani, G., and Cuschieri, A. (2011) Physical properties of carbon nanotubes for therapeutic applications. In: Carbon Nanotubes for Biomedical Applications, Klingeler, R., and Sim, R. (eds,) Springer, Germany, pp. 3-26.

75. Kumar, S., Li, B., Caceres, S., Maguire, R. G., and Zhong, W. H. (2009)

Dramatic property enhancement in polyetherimide using low-cost commercially functionalized multiwalled carbon nanotubes via a facile solution processing method. *Nanotechnology*, **20**, 465708.

76. Goh, P. S., Ng, B. C., Ismail, A. F., Aziz, M., and Sanip, S. M. (2010) Surfactant dispersed multiwalled carbon nanotube/polyetherimide nanocomposite membrane. *Solid State Sciences*, **12**, 2155-2162.

77. Kim, J. Y., Park, H. S., and Kim, S. H. (2007) Multiwall-carbon-nanotube-reinforced poly(ethylene terephthalate) nanocomposites by melt compounding. *Journal of Applied Polymer Science*, **103**, 1450-1457.

78. Ounaies, Z., Park, C., Wise, K. E., Siochi, E. J., and Harrison, J. S. (2003) Electrical properties of single wall carbon nanotube reinforced polyimide composites. *Composites Science and Technology*, **63**, 1637-1646.

79. Du, F., Fisher, J. E., and Winey, K. I. (2003) Coagulation method for preparing single-walled carbon nanotube/poly (methyl methacrylate) composites and their modulus, electrical conductivity, and thermal stability. *Journal of Polymer Science, Part B: Polymer Physics*, **41**, 3333-3338.

80. Chakraborty, A. K., and Coleman, K. S. (2008) Poly(ethylene) glycol/single-walled carbon nanotube composites. *Journal of Nanoscience and Nanotechnology*, **8**, 4013-4016.

81. Pitchan, M. K., Bhowmik, S., Balachandran, M., and Abraham, M. (2016) Effect of surface functionalization on mechanical properties and decomposition kinetics of high performance polyetherimide/MWCNT nano composites. *Composites, Part A: Applied Science and Manufacturing*, **90**, 147-160.

82. Pitchan, M. K., Bhowmik, S., Balachandran, M., and Abraham, M. (2017) Process optimization of functionalized MWCNT/polyetherimide nanocomposites for aerospace application. *Materials & Design*, **127**, 193-203.

83. Park, C., Ounaies, Z., Watson, K. A., Pawlowski, K., Lowther, S. E., Connell, J. W., Siochi, E. J., Harrison, J. S., and St. Clair, T. L. (2002) Polymer-Single Wall Carbon Nanotube Composites for Potential Spacecraft Applications. NASA, USA. Online: https://ntrs.nasa.gov/archive/nasa/casi.ntrs.nasa.gov/20030002361.pdf (assessed 15th September 2018).

84. Noll, A., and Burkhart T. (2011) Morphological characterization and modelling of electrical conductivity of multi-walled carbon nanotube/poly(p-phenylene sulfide) nanocomposites obtained by twin screw extrusion. *Composites Science and Technology*, **71**, 499-505.

85. Diez-Pascual, A. M., Guan, J., Simard, B., Gomez-Fatou, M. A. (2012) Poly(phenylene sulphide) and poly(ether ether ketone) composites reinforced with single-walled carbon nanotube buckypaper: II – Mechanical properties, electrical and thermal conductivity. *Compos-*

ites, Part A: Applied Science and Manufacturing, **43**, 1007-1015.

86. Yu, S., Wong, W. M., Hu, X., and Juay, Y. K. (2009) The characteristics of carbon nanotube-reinforced poly(phenylene sulfide) nanocomposites. *Journal of Applied Polymer Science*, **113**, 3477-3483.

87. Diez-Pascual, A. M., Naffakh, M., Marco, C., and Ellis, G. (2012) Rheological and tribological properties of carbon nanotube/thermoplastic nanocomposites incorporating inorganic fullerene-like WS_2 nanoparticles. *Journal of Physical Chemistry B*, **116**, 7959-7969.

88. Rana, S., Alagirusamy, R., and Joshi, M. (2010) Mechanical behavior of carbon nanofibre-reinforced epoxy composites. *Journal of Applied Polymer Science*, **118**, 2276-2283.

89. Rana, S., Alagirusamy, R., and Joshi, M. (2011) Development of carbon nanofibre incorporated three phase carbon/epoxy composites with enhanced mechanical, electrical and thermal properties. *Composites, Part A: Applied Science and Manufacturing*, **42**, 439-445.

90. Joshi M., and Chatterjee U. (2016) Polymer nanocomposite: An advanced material for aerospace applications. In: *Advanced Composite Materials for Aerospace Engineering*, Rana, S., and Fangueiro, R. (eds.), Woodhead Publishing, UK, pp. 241-264.

91. Chattopadhyay, D. K., and Raju, K. V. S. N. (2007) Structural engineering of polyurethane coatings for high performance applications. *Progress in Polymer Science*, **32**, 352-418.

92. Boubakri, A., Guermazi, N., Elleuch, K., and Ayedi H. F. (2010) Study of UV-aging of thermoplastic polyurethane material. *Materials Science and Engineering A*, **527**, 1649-1654.

93. Chatterjee, U., Joshi, M., and Butola B. S. (2013) Predicting Changes in TPU Coating Properties with Weathering. *Polymer Processing Society Asia/Australia Conference*, India.

94. Chatterjee, U., Joshi, M., and Butola, B. S. (2014) Weathering Performance of Polyurethane Coatings for Aerostat Applications. *International Conference on Technical Textiles and Nonwovens*, India. Online: https://www.innovationintextiles.com/international-conference-on-technical-textiles-and-nonwovens/ (assessed 15th September 2018).

95. Chatterjee, U., Joshi, M., Butola, B. S., Thakre, V., Singh, G., and Verma, M. K. (2015) Thermoplastic Polyurethane Coatings for Aerostat: Influence of Polyurethane Chemistry and Additives on Weathering Properties. *International Symposium on Polymer Science and Technology*, India.

96. Bhattacharyya, A., and Joshi, M. (2011) Co-deposition of iron and nickel on nanographite for microwave absorption through fluidized bed electrolysis. *International Journal of Nanoscience*, **10**, 1125-1130.

97. Guo, W., Jia, Y., Tian, K., Xu, Z., Jiao, J., Li, R., Wu, Y., Cao, L., and Wang, H. (2016) UV-triggered self-healing of a single robust SiO_2 microca-

psule based on cationic polymerization for potential application in aerospace coatings. *ACS Applied Materials & Interfaces*, **8**(32), 21046-21054.

PNIPAAM Grafted Porous Polymeric Monoliths: Parameters Affecting Structure and Morphology

2.1 Introduction

Solid phase synthesis and chromatographic separation technologies extensively employ polymers as supports. The polymer particles in these supports are joined together as a porous network or monolith where the surfaces of these particles provide specific reaction or adsorption sites based on charge, affinity, etc. To achieve suitable porosity in the structures, many studies have focused on the use of porogens during polymerization of styrene [1-3]. During suspension polymerization of styrene with divinylbenzene as crosslinker, the polymer chains become incompatible with the porogen and precipitate out to arrange themselves as microspheres. As the microspheres grow, they aggregate into random structures, thus, forming porous networks. Another approach called 'reactive gelation' has been reported in the literature which was observed to control the properties of the network in a better way [4]. This process included the swelling of the preformed latex particles (e.g., polystyrene particles) with monomer (e.g., styrene) followed by the particle gelation using salt and post polymerization of the swelling monomer. Attainment of both controlled morphology and porosity of the structures has been reported. The use of such a strategy is particularly beneficial in achieving the monolithic structures where the handling of the individual particles is avoided, thus, overcoming the usual agglomeration issues associated with the particles owing to their high surface energy.

Recently, a number of studies focused on grafting of functional thermo-responsive poly(N-isopropylacrylamide) (PNIPAAM) brushes from the spherical latex particles by using atom transfer radical polymerization (ATRP) of N-isopropylacrylamide [5-8]. The particles were first functionalized with a thin layer of ATRP initiator followed by grafting of PNIPAAM brushes from the surface of the

Vikas Mittal[a] and Nadejda Matsko[b]
[a]Khalifa University of Science and Technology, Abu Dhabi, UAE; [b]Graz Centre of Electron Microscopy, Graz, Austria

particles. Because of the controlled nature of atom transfer radical polymerization, the amount of grafting and hence subsequent properties of the grafted particles could be easily controlled. PNIPAAM is a material of choice because below its lower critical solution temperature (LCST) of about 32 °C, the polymer chains exhibit chain extended conformations and random coil structure and above LCST, the chains transform into more collapsed globular form. The intermolecular hydrogen bonding with the water molecules due to the chain extended morphology below LCST generates the hydrophilic nature of the chains. The intramolecular bonding between the CO and NH groups above LCST dominate over the external hydrogen bonding [9-12]. Owing to this nature of PNIPAAM, the particles modified with PNIPAAM were confirmed to hold tremendous potential for separation processes solely controlled by temperature [8]. As the free particles hold less importance than the macroporous networks or structures as chromatographic media, as a special functionalization of the monoliths, PNIPAAM chains can also be grown from the particles forming the network or monolith. This can lead to chromatographic separations of media like proteins, viruses, etc. just by changing the network or eluent temperature. This would help in avoiding the use of harsh adsorption and desorption conditions used conventionally which may sometimes affect the quality of the biological media. One important thing to note here is that the particles in these studies were prepared by surfactant free emulsion polymerization thus resulting in large particle sizes of approximately 500 nm. To generate monoliths from such particles using the earlier reported 0reactive gelation' approach would require large amounts of salt for gelation. Therefore, an alternative way is required which may overcome the use of salt during the monolith generation. One possibility is to generate loosely held aggregates from the latex particles followed by the simultaneous grafting of PNIPAAM and subsequent crosslinking of these formed PNIPAAM chains [13].

This chapter reports the synthesis of the monoliths by grafting of PNIPAAM brushes from the aggregated particles or free latex particles and simultaneously crosslinking them in order to provide the monolith structure. The PNIPAAM functionalized monoliths were prepared without the salt gelation step. Different monoliths were generated by varying the solid fraction as well as drying conditions and the effect of these variations on the resulting monolith structure and morphology was analyzed.

2.2 Experimental

2.2.1 Materials

N-isopropylacrylamide (NIPAAM, 97%) and N,N'-methylenebis(acrylamide) (MBA, ≥97%) were procured from Aldrich (Buchs, Switzerland). Styrene (S, ≥99.5%), divinylbenzene (DVB, ≈80%) and potassium peroxodisulphate (KPS, >99.0%) were purchased from Fluka (Buchs, Switzerland) and were used as supplied without further purifications. Atom transfer radical polymerization initiator end capped with an acrylic moiety (2-(2-bromopropionyloxy) ethyl acrylate, BPOEA) was synthesized as reported earlier [14].

Reagents to run the ATRP polymerization, namely 1,1,4,7,10,10-hexamethyltriethylenetetramine (HMTETA, 97%), copper(I) bromide (CuBr, 99.99%), copper(II) bromide (CuBr$_2$, 99.99%) and powder copper (Cu, 99%, 200 mesh), were procured from Aldrich (Buchs, Switzerland). Ultra-pure Millipore water was employed in all experiments.

2.2.2 Synthesis of Polystyrene Particles and Surface Functionalization

Polystyrene seed latex was prepared by surfactant free emulsion polymerization of styrene (14 g), divinylbenzene (0.5 g) added with 0.3 g KPS and 310 g of water [7]. The crosslinked polystyrene latex particles were washed by repeated centrifugation and resuspension cycles in millipore water and were subsequently functionalized with a thin layer of ATRP initiator on the surfaces. For this, BPOEA (0.21 g), DVB (0.065 g) and styrene (0.26 g) (0.01 g of KPS in 0.5 mL of water) were added to15 g of preformed latex heated to 70 °C and polymerized for further 7 h [7]. Finally, the functionalized latex was washed following the similar process of centrifugation and resuspension in water.

2.2.3 Shear Aggregation

The functionalized latex particles were sonicated for 10 min (ultrasonic horn at 70% amplitude) and shear mixed for another 10 min with a ultra-high shear mixer (Ultra-Turax T50) to destabilize the latex. The resulting porous latex aggregates were allowed to settle and the aqueous layer was decanted to concentrate the latex solid fraction.

2.2.4 Monolith Synthesis

Different monoliths were prepared from these concentrated aggregates as well as concentrated functionalized latex particles. For monoliths prepared from aggregates, the concentrated latex (1.5 g, 15 wt%) was transferred to a flat bottom vial, degassed and subsequently added with HMTETA (17 mg), CuBr (4.3 mg), $CuBr_2$ (1.4 mg) and Cu (2 mg). The solution was stirred for 2 min and NIPAAM (0.4 g) and MBA (0.4 g) were added to the solution. The viscous slurry was vigorously stirred to solubilize the monomers. The stirrer was then removed, and the vial was degassed and kept for 12 h without shaking. In another trial, exactly same process and reagents were used but, in this case, additional 1.5 mL of water was added to the slurry in order to see the effect of volume changes on the properties of the monolith.

The formed monoliths were carefully removed from the vials and dried at room temperature. Drying of monoliths at 70 °C was also carried out in order to analyze the effect of fast drying conditions on the structure of the monolith. ATRP initiator modified particle latex was also concentrated by evaporating a fraction of water followed by monolith synthesis following the similar procedure as reported above. In addition, the effect of different solid fractions on the structure and morphology of the resulting porous monoliths was also analyzed.

2.2.5 Electron Microscopy

Hitachi field emission in-lens S-900 high resolution scanning electron microscope (SEM) at accelerating voltages of 10-20 kV was employed to observe the surface morphology of the particles. Carbon/collodium coated 400-mesh copper grids, freshly etched by charged oxygen plasma (10 s, 100 mV, 5 mbar of O_2) in Balzers GEA-003-S glow-discharge apparatus (Balzers), were placed on the droplets of particle suspensions for 2 min, dried on filter paper followed by sputter coating with 3 nm platinum.

The SEM of monoliths was performed by fixing small pieces of dry monoliths on copper supports followed by sputter coating with 3 nm platinum.

2.3 Results and Discussion

PNIPAAM modified networks hold tremendous potential for their use in commercial separation technologies [8,13]. PNIPAAM grafting of particles along with simultaneous monolith generation presents an alternative approach to the earlier reported reactive gelation process in order to obtain such commercially important materials. The potential of these PNIPAAM grafted particles as well as monoliths for such separations has been quantitatively established [8, 13]. The temperature dependent swelling deswelling studies on the monoliths as well as adsorption desorption of a virus sample on the PNIPAAM grafted particles was successfully reported. But an optimum porous structure and morphology is required in the porous networks for the process to be successful. Few preliminary trials in this regard to analyze the effect of various process changes on the resulting monolith network structure and morphology are detailed as follows.

Surfactant free polymerization of polystyrene particles in the presence of small amount of crosslinker led to the generation of narrowly dispersed latex particles. These particles were subsequently covered with a thin shell (80 nm thick) of ATRP initiator (BPOEA). The terminal acrylic group of BPOEA molecule could be copolymerized with styrene and divinylbenzene in order to obtain this thin layer of ATRP initiator [7]. The copolymer chains were observed to be compatible with the surface of the polystyrene particles and no secondary nucleation of the copolymer particles was observed as the increase in the diameter of the seed particles after forming a shell is quantitative to the amount of monomers initially added. The amount of crosslinker can be altered in order to change the density of ATRP initiator in the shell which subsequently affects the density of PNIAPAAM brushes grafted from the surface of particles. The grafting of ATRP initiator on the particle surface was also quantitatively confirmed by NMR [7]. The SEM image of these functionalized particles is shown in Figure 2.1a. The particles are still fairly monodisperse though sedimentation sets in with time owing to their synthesis without the emulsifier. The average hydrodynamic diameter of the particles was estimated by laser light scattering to be 530 nm. When these latex particles were subjected to high shear forces in a sonicator and shear mixer, a network of physical aggregates resulted owing to the destabilization and collapse of the latex. Figure 1b shows the SEM image of these aggregates. The aggregates are clearly visible to be porous thus providing the ideal material to generate monoliths from them. The average size of these networks was estimated be 3-4 µm. However, as the forces holding these particles together in these aggregates are only physical, therefore, these aggregates are not useful for commercial application and much stronger bonding forces are required to preserve the so formed porous structure.

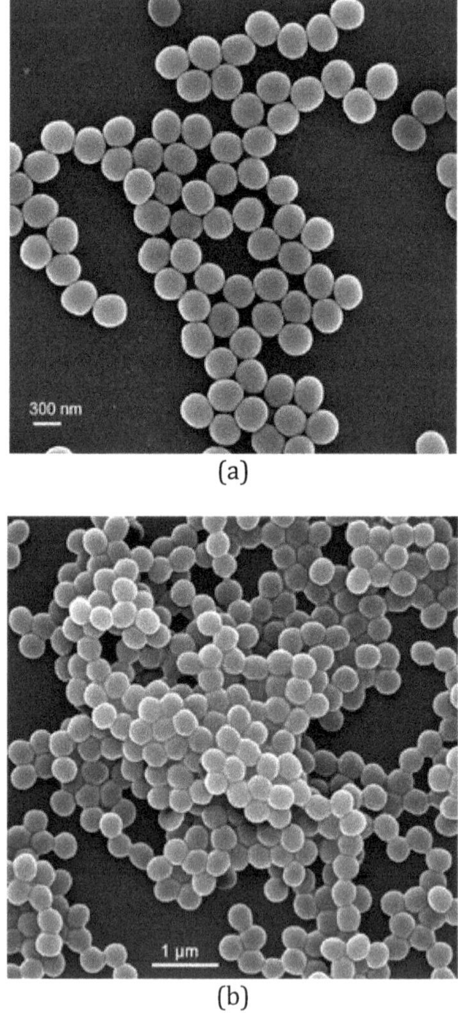

(a)

(b)

Figure 2.1 (a) SEM micrograph of ATRP initiator functionalized polystyrene particles and (b) SEM micrograph depicting these particles when aggregated using high shear.

Atom transfer radical polymerization of NIPAAM from the aggregated particles led to the successful generation of monoliths. Use of high extents of crosslinker helped to attain the required stability of the monoliths. The monoliths were hard when totally dry, whereas they turned soft and swollen when wet. Monoliths prepared by low and high solid fraction were both stable under these conditions. However, very different morphologies were observed under SEM for these two kinds of monoliths. Figures 2.2a, b show the SEM images of the monolith with higher solid fraction used. Surprisingly, the monolith had a number of channels or tubes clearly visible in its morphology. These channels had random dimensions and were scattered pointing in all directions. Another thing to note here is that these tubes were very smooth from inside and had particles adsorbed all around their external surface. Figures 2.2c, d depict the morphology of the monolith which was prepared at half the solid content of the monolith shown in Figures 2.2a, b. Almost no or very negligible amount of earlier seen channels or tubes could be observed in this case and the particles were nicely connected forming a homogenous porous structures (Figure 2.2d). Figures 2.2e, f show the channels seen in Figures 2.2a, b in more detail. Absence of so formed channels in the second monolith indicates that the solid fraction may be responsible for the formation of these channels, as there was no other difference in the generation of the two monoliths. It was observed that the aggregates quickly turned solid as soon as NIPAAM and MBA solution was mixed with them thus quickly trapping all the water present in the system. It is suspected that while drying the water trapped in the system has more difficulty to escape in the monolith which is much more dense. Thus, the so formed channels formed in the former monolith can be explained to be the trails of the trapped water molecules shooting out of the structure making their path where ever they could find the way. Although water could have escaped through the pores, but the water trapped in the areas of high concentration of formed PNIPAAM polymer led to generation of these tubes with smooth inside surfaces as the polymer may be swollen in the presence of the trapped water. However, the monolith with lower solid content seem to have more porosity and subsequently less concentration of any dead areas containing only polymer thus easing the escape of water from the monolith structure. This indicates that for the operating conditions used for monolith generation, the critical solid fraction should not be exceeded to preserve the homogenous morphology of the monolith.

 Functional Polymer and Composite Systems

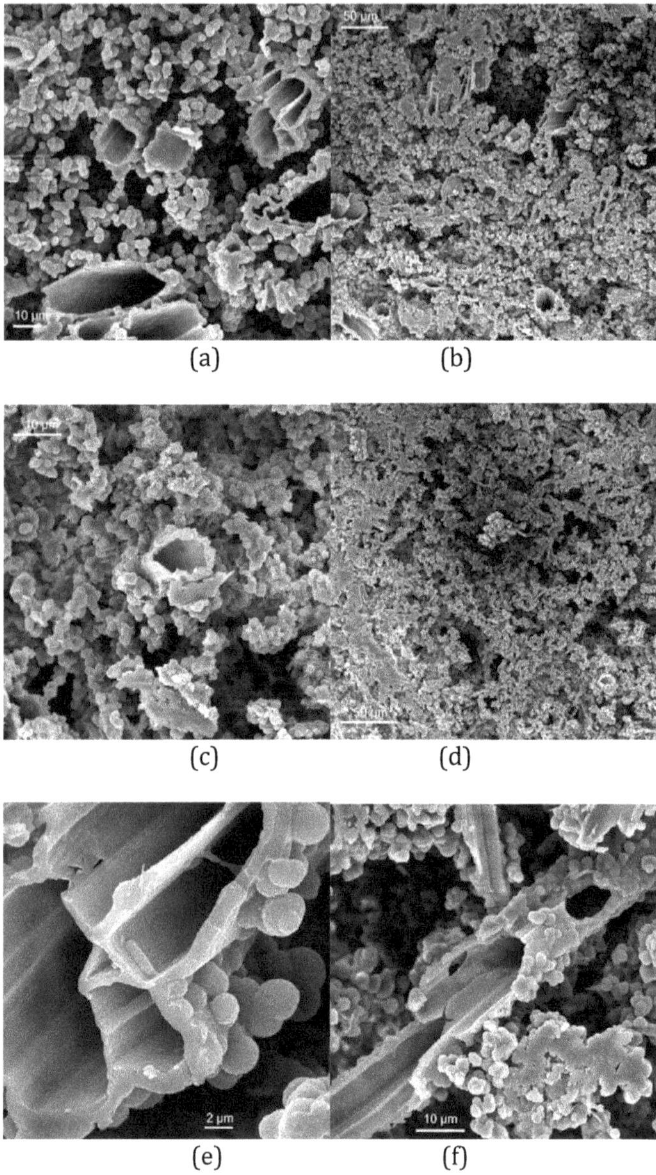

(a)   (b)

(c)   (d)

(e)   (f)

Figure 2.2 (a) & (b) High and low magnification scanning electron micrographs of the monolith generated from aggregated particles using higher solid fraction and dried at room temperature, (c) & (d) high and low magnification SEM micrographs of the monolith prepared using lower solid fraction and dried at room temperature and (e) & (f) scanning micrographs representing the tubes or channels in the structures, as seen in the Figures 2.1a and 2.1b.

Effect of drying at higher temperatures was analyzed in order to establish optimum drying conditions. Figure 2.3 shows the SEM images of the monolith similar to shown in Figures 2.2a, b, but dried at

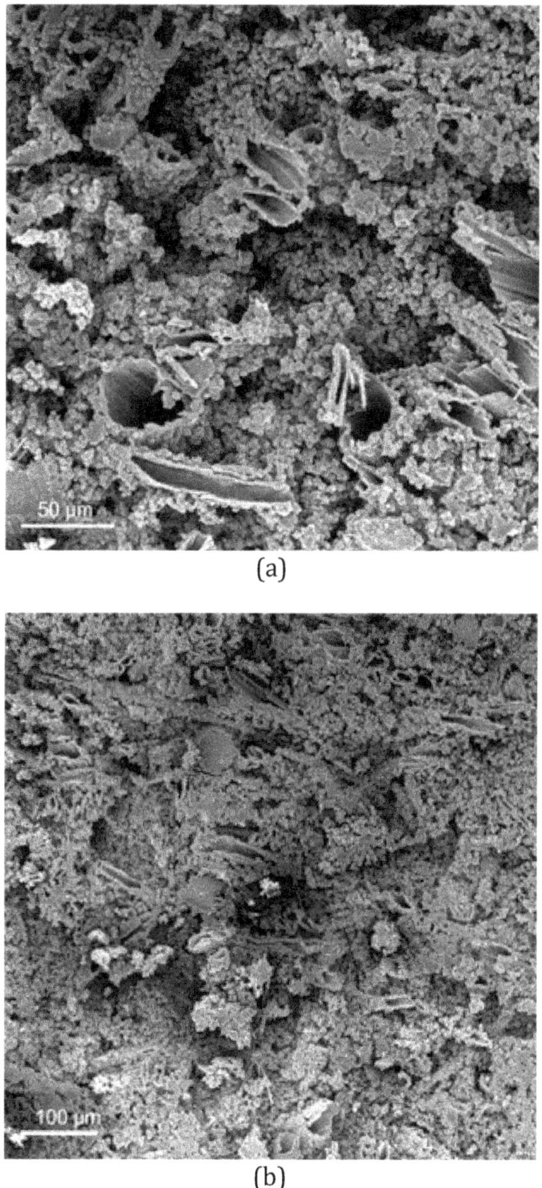

(a)

(b)

Figure 2.3 (a) and (b) High and low magnification images of the monoliths prepared using higher solid fraction and dried at 70°C.

70 °C. On comparison of the images of monoliths dried at room temperature and 70 °C, it is clearly evident that the number of channels or tubes significantly increased when the monolith was dried at 70 °C. It also supports the earlier hypothesis concerning the escape of trapped water from the monolith structure. Heating at 70 °C is expected to accelerate this escape tremendously, thus forcing all of the water molecule to escape collectively in a much shorter period of time. This definitely is expected to increase the number of these channels or pathways through which the water molecules tend to escape. Thus it points to an another important observation to bear in mind that drying conditions have very significant effect on the final structure of the monolith and care should be taken to use optimal drying conditions.

Similar effect was also observed when the concentrated latex particles (not aggregated) were themselves grafted with PNIPAAM and simultaneous crosslinking with MBA. The particles were grafted and networked in order to analyze the effect of starting media on the final structure of the monolith. Similar tube or channel structures were observed when higher solid fractions were used (Figure 2.4a), though the number of these channels were relatively smaller than those seen in the case of high solid fraction monoliths obtained from aggregated particles. This can be the result of easy escape of water from the latex particle monoliths owing to the absence of any physical network in the beginning to trap the water of the system. The cross-section of the channel structure was also similar as shown in Figure 2.4b, with smooth inside surface and particles adsorbed on the outer walls. The monolith generated with lower solid content was observed to be totally free from any of the channels or tubes owing to the similar reasons mentioned above (Figure 2.4c).

The obtained results, thus, indicate that successful generation of the monoliths could be achieved by aggregate or particle grafting process and various factors affecting the final stricture of the monolith could be established. In order to obtain homogenous and porous network, thus to avoid the tubes or channels in the structure, optimum values of solid fraction and optimal drying procedures are required. Another point to note here is that in the conditions used to generate the monoliths, higher amounts of crosslinker was used. However, this amount can also be varied according to the requirement of the process thus indicating controlled fine-tuning of the process. Apart from that, density of the ATRP initiator on the surface of the functionalized particles can also be easily controlled according to the requirement of the process. The formed monoliths represent the separation supports which have high potential for use in adsorption/desorption processes solely assisted by temperature.

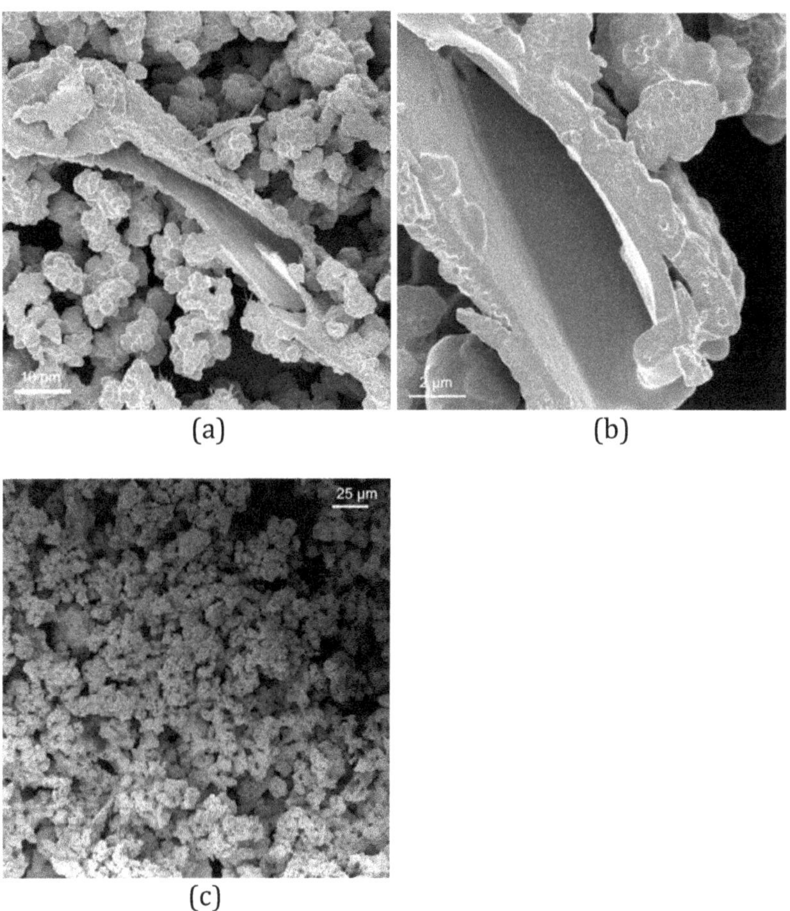

(a)

(b)

(c)

Figure 2.4 (a) SEM micrograph of the monolith prepared from latex particles using high solid fraction and dried at room temperature, (b) high magnification image of the channel structure observed in Figure 2.4a and (c) SEM image of the monolith prepared from latex particles using lower solid fraction and dried at room temperature.

2.4 Conclusions

Shear mixing of the latex particles led to destabilization and collapse of the emulsifier-free latex particles into physical networks of the aggregated particles. These aggregates and particles could successfully be transformed into solid porous monoliths by simultaneous grafting and crosslinking of PNIPAAM from the surface. Various factors affecting the structure and morphology of these monoliths could be analyzed. Under the operating conditions used to generate the monoliths, it was found that after a critical solid fraction of the aggregates or particles used to form the monolith, a surprising channel or tube morphology in the monolith structure was developed. This channel or tube morphology was observed to be enhanced when the monoliths were dried at higher temperatures especially in the monoliths with higher solid fractions. The monolith structure can easily be fine-tuned or altered according to the requirement by changing the amount of crosslinker and the amount of ATRP initiator on the surface of the particles thus indicating the robustness of the technique.

Summary of the Results

Crosslinked polystyrene latex particles generated with surfactant free emulsion polymerization were functionalized with thin layer of ATRP initiator on the surface. The functionalized particles were aggregated as physical networks using high shear techniques. The latex particles as well as the aggregated particles were used to successfully generate thermo-responsive monoliths by grafting PNIPAAM brushes from the surface of the particles by atom transfer radical polymerization and simultaneous crosslinking of these brushes. Different solid fractions and drying conditions were used in the generation of the monoliths and the effect of these variations on the structure and morphology of the monoliths was analyzed. Surprising morphologies consisting of tubes or channels were observed when higher solid fractions were used. This behavior was enhanced when the monoliths were dried at higher temperatures. By using proper solid fractions and drying conditions, homogenous porous monoliths could be achieved.

Acknowledgement

The final version of the chapter has been earlier published in Journal of Porous Materials, 2009, 16:537-543. Reproduced with permission from Springer Nature.

References

1. Okay, O. (2000) Macroporous copolymer networks. *Progress in Polymer Science*, **25**, 711-779.
2. Guyot, A., and Bartholin, M. (1982) Design and properties of polymers as materials for fine chemistry. *Progress in Polymer Science*, **8**, 277-331.
3. Seidl, J., Malinsky, J., Dusek, K., and Heitz, W. (1967) Makroporöse Styrol-Divinylbenzol-Copolymere und ihre Verwendung in der Chromatographie und zur Darstellung von Ionenaustauschern. *Advances in Polymer Science*, **5**(2), 113-123.
4. Marti, N., Quattrini, F., Butte, A., and Morbidelli, M. (2005) Production of polymeric materials with controlled pore structure: the "reactive gelation" process. *Macromolecular Materials and Engineering*, **290**, 221-229 (2005).
5. Kizhakkedathu, J. N., Takacs-Cox, A., and Brooks, D.E. (2002) Synthesis and characterization of polymer brushes of poly(N,N-dimethylacrylamide) from polystyrene latex by aqueous atom transfer radical polymerization. *Macromolecules*, **35**, 4247-4257.
6. Kizhakkedathu, J. N., and Brooks, D. E. (2003) Synthesis of Poly(N,N-dimethylacrylamide) Brushes from charged polymeric surfaces by aqueous ATRP: Effect of surface initiator concentration. *Macromolecules*, **36**, 591-598.
7. Mittal, V., Matsko, N. B., Butte, A., and Morbidelli, M. (2007) Functionalized polystyrene latex particles as substrates for ATRP: Surface and colloidal characterization. *Polymer*, **48**, 2806-2817.
8. Mittal, V., Matsko, N. B., Butte, A., and Morbidelli, M. (2007) Synthesis of temperature responsive polymer brushes from polystyrene latex particles functionalized with ATRP initiator. *European Polymer Journal*, **43**, 4868-4881.
9. Ringsdorf, H., Sackmann, E., Simon, J., and Winnik F. M. (1993) Interactions of liposomes and hydrophobically-modified poly-(N-isopropylacrylamides): An attempt to model the cytoskeleton. *Biochimica et Biophysica Acta - Biomembranes*, **1153**, 335-344.
10. Bae, Y. H., Okano, T., and Kim, S. W. (1990) Temperature dependence of swelling of crosslinked poly(N,N′-alkyl substituted acrylamides) in water. *Journal of Polymer Science, Part B: Polymer Physics*, **28**, 923-936.
11. Heskins, M., and Guillet, J. E. (1968) Solution properties of poly(N-isopropylacrylamide). *Journal of Macromolecular Science, Part A: Chemistry*, **2**, 1441-1455.
12. Yoshida, R., Uchida, K., Kaneko, Y., Sakai, K., Kikuchi, A., Sakurai, Y., and Okano. T. (1995) Comb-type grafted hydrogels with rapid deswelling response to temperature changes. *Nature*, **374**, 240-242.
13. Mittal, V., Matsko, N. B., Butte, A., and Morbidelli, M. (2008) Swell-

ing deswelling behavior of PS-PNIPAAM copolymer particles and PNIPAAM brushes grafted from polystyrene particles & monoliths. *Macromolecular Materials and Engineering*, **293**, 491-502.

14. Matyajaszewski, K., Gaynor, S. G., Kulfan, A., and Podwika, M. (1997) Preparation of hyperbranched polyacrylates by atom transfer radical polymerization. 1. Acrylic AB* monomers in "living" radical polymerizations. *Macromolecules*, **30**, 5192-5194.

3

Polymer Nanocomposites for Automotive Applications

3.1 Introduction

Nanomaterials are dominant component of scientific research and development owing to their outstanding performance and ability to significantly strengthen the performance of other materials. Polymer nanocomposites are emerging class of high performance nanomaterials which are useful for a myriad of applications [1]. For automotive applications, polymeric materials offer many benefits like lightweight, ease of processing, appreciable strength, corrosion resistance and cost effectiveness in terms of fabrication and recycling over metallic materials. These benefits of polymers lead to weight reduction, energy effectiveness, low exhaust emissions, safety and performance in automotive applications [2,3]. In order to further strengthen the performance of polymers and, thereby, to widen their automotive practicability, polymer modification by means of nanocomposite generation has evolved as a common strategy. Interestingly, the incorporation of a small amount of nanomaterials is observed to be beneficial to significantly enhance concurrent improvement in stiffness and toughness, thermal stability, flame retardancy, electrical and electronic features, tribological features, oil and water impermeability, chemical stability, etc., as compared to micromaterials. Such effects in polymer nanocomposites occur due to much higher interfacial surface area of the filler nanoparticles and corresponding polymer-filler interfacial interactions [4-12]. As a result, high performance polymer nanocomposites are useful to generate high fuel efficiency, minimal exhaust emissions as well as environmental impact, strength, noise reduction, automotive body parts with self-healing capability, automobile windshield, automotive coatings and paints, electrical and electronic components, breaking system and shock absorbers, tires, inner tubes, etc. [4,6,13-16].

The first successful commercial automotive application of polymer nanocomposites was reported by researchers at Toyota Motor

Anish M. Varghese and Vikas Mittal, Khalifa University of Science and Technology, Abu Dhabi, UAE

Corporation, Japan through the development of nylon 6/montmorillonite (MMT) nanocomposite as an effective car engine cambelt cover possessing superior heat resistance. In this system, the use of 4.2 wt% of MMT was beneficial to strengthen the thermal and mechanical characteristics of nylon 6 [4,17,18]. Later in 2002, General Motors Company added another important milestone for the use of polymer nanocomposites in automotive applications, through the utilization of 3 wt% exfoliated nano-clay reinforced thermoplastic polyolefin composite for effective external light-weight step assist applications [7,19]. An interesting research carried out by Chevrolet reported about 25% reduction in weight for nano-clay reinforced polymer based automotive body side trim [20]. In the same year, Pirelli introduced winter tires by mixing nanoscale carbon black with rubber [21]. Afterwards, polymer nanocomposite based cargo bed from General Motors Company, under-hood cover for timing gauge from Toyota Motor Corporation and engine cover from Mitsubishi Motors Corporation were commercialized [22,23]. With time, a large number of developments related to polymer nanocomposite based automotive applications have been generated from both industry and academia. In this respect, the automotive applicability of different nanofiller reinforced polymers representing the family of thermoplastics and thermosets has been evaluated to obtain enhanced performance and weight saving together with acceptable cost to performance ratio. Among various polymer nanocomposites developed over time, some are already been commercialized for diverse automotive applications with varying degrees of success, while many are still in various stages of development and commercialization, thus, are expected to be available in the market over time.

The current review aims to summarize the recent developments with respect to automotive applications of polymer nanocomposites.

3.2 Polymer Nanocomposites for Automotive Tires

Automotive tire is an inflatable toroidal elastomer based composite structure which envelops the wheel rim to provide good vehicle traction under wet and dry road conditions. It is also considered as a most prominent automotive segment from the family of polymeric materials and should offer better balance between three important variables such as traction, rolling resistance and abrasion resistance in order to provide superior performance and service life. Tread portion of the tire has a significant influence on tire functioning, thus, much attention has been focused to enhance tire tread performance and, thus, durability by means of nanomaterial incorporation [23-25].

3.2.1 Styrene Butadiene Rubber (SBR) Nanocomposites

Li *et al.* [26] reported the development of SBR/magadiite nanocomposite based tire tread exhibiting benefits in terms of weight saving and performance. The utilization of cetyltrimethylammonium (CTMA) functionalization of magadiite was observed to be beneficial in improving its interaction with SBR and corresponding reinforcing capability. As a result, the resultant nanocomposite system exhibited higher mechanical performance compared to same amount of silica reinforced SBR, which confirmed its potential for use in automotive energy effectiveness applications [26]. In another study, Mao *et al.* [27] reported that silane-mediated SBR-magadiite grafts within the magadiite interlayer space are necessary to achieve mechanical reinforcement in the nanocomposites. Mao *et al.* [28] also compared the effect of three inorganic nanofillers namely MMT, magadiite and precipitated silica on the performance of SBR for ascertaining effective automotive tire tread applications. Among various nanocomposite systems, CTMA functionalized MMT filled SBR exhibited better storage modulus under the studied temperature range, along with higher tangent and Young's modulus owing to the formation of network morphology resulting from efficient exfoliation of nano-MMT sheets. Overall, in order to obtain weight saving benefit and subsequent energy effectiveness, the authors recommended the usage of CTMA functionalized MMT or dodecylpyridinium functionalized magadiite in the SBR based tire tread composition.

In order to develop automotive tire materials based on SBR nanocomposites, Kim *et al.* [29] studied the effect of silanized as well as aminated nano-MMT in the presence and absence of a plasticizer, poly(ethylene glycol) (PEG). Superior wet grip and reduced rolling resistance were observed for PEG-free SBR/silica/carbon black/aminated-MMT nanocomposite vulcanizate. At the same time, modulus and T_g values were higher for PEG containing SBR/silica/carbon black/silanized-MMT nanocomposite vulcanizate. In another study, Kim *et al.* [30] detailed the development of high performance passenger car tire tread using aminated nano-MMT/silica/carbon black hybrid filler reinforced SBR nanocomposites together with calcium stearate lubricant in the compound formulation. Effective static and dynamic mechanical properties, along with resistance to abrasion were noted for the nanocomposite vulcanizate.

Song [31] reported the usefulness of combinatorial filler system based on CTMA functionalized multi-walled carbon nanotubes (MWCNTs) and hydroxylated MMT nanoplatelets to upgrade the properties of SBR based tire tread formulation. Due to the synergistic effect and optimal filler dispersion, the use of combinatorial filler system in the SBR based tire tread composition exhibited significantly improved gas impermeability (9.16×10^{-45} m^4 s^{-1} N^{-1}) as well as thermal (0.3779 W m^{-1} K^{-1}) and mechanical properties (such as modulus (0.31 MPa) and tensile strength (75 MPa)) at low loadings. In addition, superior performance such as excellent wet grip, improved wear resistance and reduced rolling resistance were observed for pneumatic tire generated using the SBR nanocomposite.

Tang *et al.* [32] examined the influence of nano-MoS_2 platelets incorporation on the carbon black dispersion in the SBR matrix. The use of nano-MoS_2 platelets was advantageous in diminishing the degree of carbon black flocculation in the SBR matrix and, thereby, improving filler dispersion. The resultant MoS_2/carbon black hybrid reinforced SBR displayed 50% increase in the tensile modulus and 10 °C decrease in the heat build-up with 3 phr of MoS_2.

In order to obtain both economic and environmental benefits during automotive tire manufacturing, Roy *et al.* [33] reported the application of nano-ZnO, instead of conventional ZnO, in SBR based tire formulation. The authors studied the effect of surface functionalization of nano-ZnO on its dispersion state in the SBR matrix and, thus, the resultant nanocomposite performance. The thermal stability, mechanical properties and crosslink density of SBR were appreciably enhanced on incorporating 1 phr silanized nano-ZnO compared to SBR nanocomposite with 5 phr conventional ZnO, attributed to the homogeneous filler dispersion and strong chemical interaction with SBR. In another study, Thaptong *et al.* [34] compared the effect of conventional as well as highly dispersible silica nanoparticles on the performance of SBR based tire tread. The use of both types of silica nanoparticles had almost similar effect on the wet grip as well as abrasion and rolling resistances of SBR tire tread. Also, almost comparable modulus and heat build-up were observed for the two SBR/silica nanocomposites. The influence of silanization temperature on the SBR nanocomposite properties was also investigated and increasing tendency with enhancement in silanization temperature was observed.

Qiao *et al.* [35] reported the usefulness of epoxidized SBR/silica nanocomposites for automotive tire applications. The introduction of epoxy groups into SBR was advantageous in obtaining improved dispersion of silica nanoparticles and, thus, enhanced interfacial interactions through the formation of covalent bonds without additional coupling agents. Overall, the resultant epoxidized SBR/silica nanocomposites demonstrated improved mechanical performance, lower rolling resistance and better wet grip. The tensile strength was observed to be highest at 29.4 MPa at a glycidyl methacrylate (GMA) GMA content of 4.8 wt%, which was employed as the epoxy group-included monomer.

With an aim to design green automotive tire tread material, Sun *et al.* [36] discussed the adequacy of nano-SiO_2/microcrystalline cellulose hybrid filler reinforced SBR/silica composite. The introduction of nano-SiO_2/microcrystalline cellulose hybrid filler into silica filled SBR enhanced static and dynamic mechanical properties owing to the enhanced polymer-filler interactions. Compared to SBR/silica composite and its microcrystalline cellulose filled version, hybrid filler reinforced SBR/silica composite based tire tread vulcanizate exhibited a superior wet grip and an ultra-low rolling resistance.

Saeed *et al.* [37] disclosed the usefulness of SBR/polybutadiene (PB)/silane functionalized silica nanocomposites in the tire tread formulation of passenger cars. After curing in the presence of activator system based on non-sulfur donor and ZnO, the resultant nanocomposites exhibited outstanding performance such as excellent mechanical properties, low heat build-up, superior abrasion resistance and substantial low temperature tan δ value which indicated desirable grip under wet and ice conditions. Overall, the amount of the chemical curatives needed for tire tread were reduced by approximately 58% by weight. In another study, Ostad-Movahed *et al.* [38] reported the introduction of silanized nano-silica into SBR/PB blend based passenger car tire formulation. The addition of nano-silica was advantageous in upgrading the static and dynamic mechanical properties and abrasion resistance of SBR/PB blend.

Veiga *et al.* [39] prepared truck tire tread vulcanizate possessing ultra-low rolling resistance (decreased by 10%), wet grip (increased by 18.5%) and superior mechanical properties based on nano-carbon black/silica hybrid filler reinforced SBR/natural rubber (NR)/PB blend nanocomposites. Optimal performance was observed for the nanocomposite containing 35 phr nano-carbon black, 15 phr silica and 3 phr silane coupling agent.

In a recent study, Malas *et al.* [40] investigated the role of expanded graphite and its oxidized version on the properties SBR in order to generate effective automotive tire tread materials. The presence of both unmodified and modified fillers had positive effect on the thermo-mechanical properties and abrasion resistance of SBR, which were higher for the composite based on oxidized expanded graphite due to sizeable interlayer spacing and presence of large number of surface functional groups. Furthermore, the incorporation of nanofillers into carbon black containing SBR was beneficial to improve the storage modulus even at sub-ambient temperature. In order to enhance the thermal transportation capability and mechanical features of SBR for durable automobile tire applications, Yin *et al.* [41] also made use of ionic liquid-modified graphene oxide (GO). Homogeneous dispersion of ionic liquid-modified GO within the SBR matrix was observed. The tensile strength, tear strength, thermal conductivity and solvent resistance of the nanocomposite with 5 phr filler content were observed to enhance by 505, 362, 34 and 31%, respectively, as compared to the pure polymer.

During the service life of a tire, hysteresis is one of the most important parameters which determines the energy loss through it. For elastomeric nanocomposites, hysteresis loss depends largely on the dispersion of nanofillers and degree of interfacial interaction with elastomers. In this respect, Yang *et al.* [42] reported the generation of a strong covalent interface between SBR and graphene through *ortho*-quinone functionalization of graphene. Owing to the effective filler dispersion and optimal interaction with SBR, the nanocomposites demonstrated highly reduced dynamic energy loss. Also, top most energy efficiency of "A grade" was observed for automotive tire designed using the developed nanocomposites.

Table 3.1 summarizes the effect of various nanomaterials on the performance of SBR based automotive tire vulcanizate materials.

3.2.2 Natural Rubber Nanocomposites

Rooj *et al.* [43] observed a remarkable improvement in fatigue crack growth resistance for NR/carbon black composites on incorporating

Table 3.1 Effect of various nanomaterials on the performance of SBR based automotive tire vulcanizate materials

Reference	Nanomaterials	Achievements
Li *et al.* [26]	Cetyltrimethylammonium-modified magadiite	Increase in mechanical properties
Mao *et al.* [28]	Cetyltrimethylammonium-modified MMT	Increase in storage modulus, tangent modulus and Young's modulus
Kim *et al.* [29]	Aminated MMT	Increase in wet grip, reduction in rolling resistance.
Song *et al.* [31]	Cetyltrimethylammonium-modified MWCNTs/hydroxylated MMT nanoplatelets hybrid	Increase in gas impermeability, thermal properties, mechanical properties, wet grip and wear resistance. Reduction in rolling resistance
Tang *et al.* [32]	MoS_2 nanoplatelets	Increase in static and dynamic mechanical properties. Reduction in heat build-up.
Roy *et al.* [33]	Silanized nano-ZnO	Increase in thermal stability, mechanical properties and crosslink density.
Sun *et al.* [36]	Nano-SiO_2/microcrystalline cellulose hybrid	Increase in static and dynamic mechanical properties.
Saeed *et al.* [37], Ostad-Movahed *et al.* [38]	Silanized nano-silica	Increase in static and dynamic mechanical properties, abrasion resistance and wet grip. Reduction in heat build-up.
Veiga *et al.* [39]	Nano-carbon black/silica hybrid	Increase in wet grip, and mechanical properties. Reduction in rolling resistance.
Malas *et al.* [40]	Expanded graphite and its oxidized version	Increase in thermo-mechanical properties and abrasion resistance
Yin *et al.* [41]	Ionic liquid-modified GO	Increase in thermal conductivity, tear strength, tensile strength and solvent resistance.
Yang *et al.* [42]	*ortho*-quinone-modified graphene	Reduction in dynamic energy loss.

5 wt% expanded organo-MMT nanoparticles, which confirmed the potential for tire engineering application. The authors also reported

a successful substitution of about 40% of carbon black in the nanocomposite formulation by expanded organo-MMT nanoparticles without any negative effect on the mechanical properties.

To strengthen the gas permeability of NR for engineering tire applications, especially for radial tires, Battacharaya *et al.* [44] reported the use of nanomaterials namely Cloisite 15A, sepiolite and carbon nanofiber. On incorporating the nanomaterials into carbon black filled NR, the oxygen impermeability was observed to increase due to enhanced interfacial interactions and resultant reduction in molecular mobility resulting from reduced free volume. Among the three nanocomposites, the extent of increase in oxygen impermeability was higher for Cloisite 15A nanocomposite, attributed to the influence of more tortuous path.

For enabling superior tire engineering applications, Kim *et al.* [45] described the suitability of silica/organo-MMT hybrid filler reinforced NR nanocomposite vulcanizate. The use of 7 phr *N,N*-dimethyldodecylamine-modified MMT in the composite formulation was beneficial to enrich the dispersion state of silica in the NR matrix and, thus, to form strong filler-elastomer crosslinks. Enhanced static and dynamic mechanical properties and glass transition temperature (T_g) were observed for the nanocomposite vulcanizate. Furthermore, the observed skid and rolling resistance for the NR/silica/organo-MMT nanocomposite tread vulcanizate were more desirable as compared to NR/silica composite.

Shan *et al.* [46] reported significant enhancement in mechanical properties and abrasion resistance of NR/SBR based tire tread formulation after the incorporation of 3 phr organo-MMT due to filler exfoliation and homogeneous dispersion within the NR/SBR blend. The tensile and tear strength of the nanocomposite with 3 phr organoclay were observed to enhance by 92.8% and 63.4%, respectively. In addition, the organo-MMT resulted in reduced scorch and cure time due to its capability to act as curing accelerator. In another study, Pal *et al.* [47] obtained enhanced abrasion resistance and mechanical properties for a truck tire tread formulation by means of epoxidized NR (ENR)/organo-MMT nanocomposite addition into NR/SBR blend. The rubber compound containing 70 wt% of NR, 30 wt% of SBR and 10 wt% of ENR/nano-clay was observed to be the toughest rubber.

Tavakoli *et al.* [48] prepared maleic anhydride-*grafted*-ethylene-propylene-diene monomer rubber (EPDM-*g*-MA) and ENR compatibilized NR/SBR/organo-clay nanocomposites with an objective to achieve optimized tire performance. EPDM-*g*-MA compatibilized nanocomposites exhibited higher extent of enhancement in static and dynamic mechanical properties as well as fatigue life. Zarei *et al.* [49] reported the potential of NR/PB/SBR blend/clay nanocomposites for effective automotive tire tread application. Nano-clay Cloisite 15A enhanced mechanical, rheological and curing properties of NR/PB/SBR blend, attributed to the effective elastomer intercalation into the nano-clay galleries. The ternary blend nanocomposite vulcanizate exhibited an effective improvement in wet skid resistance and a decrease in heat build-up. Wang *et al.* [50] prepared NR/ENR/silica nanocomposites with the help of wet masterbatch approach and reported their potential for utilization as superior automotive tire tread compounds. In comparison to conventional dry mixing approach, nanocomposite prepared using wet masterbatch approach displayed superior nano-silica dispersion and interfacial bonding, which resulted in improved static and dynamic mechanical properties, and reduced rolling resistance. Moreover, reduced curing time and increased crosslink density were observed in the case of wet masterbatch generated nanocomposite.

Utara *et al.* [51] reported the applicability of azobisisobutyronitrile-functionalized nano-SiC reinforced ENR/NR blend nanocomposites for the generation of effective summer automotive tires. In the study, the use of functionalized nano-SiC was helpful to obtain uniform dispersion in the elastomer matrix. The nanocomposites demonstrated improved static and dynamic mechanical properties, crosslink density and wet grip. Also, heat build-up and compression set of ENR/NR blend were diminished after the incorporation of nano-SiC. The performance of nanocomposites with neat SiC remained lower than the nanocomposites with functionalized filler due to poor filler dispersion and weak interfacial interactions.

In order to reduce the negative environmental impacts and toxicity issues related to conventional tire compound formulations, Manoharan and Naskar [52] developed a new formulation comprising terpene-modified NR, ENR, dispersible silica, nano-ZnO and other ingredients, excluding conventional ZnO activator, processing oil and silane coupling agent. The developed tire compound displayed a superior Payne effect and remarkably enhanced dynamic mechanical performance owing to uniform silica dispersion and optimal interfacial interactions. In addition, significantly reduced rolling resistance was observed for the tire compound from the observed tan δ values.

Fu *et al.* [53] detailed the development of a new automotive tire tread compound using NR/carbon black/organo-Al$_2$O$_3$ nanocomposites. The nano-Al$_2$O$_3$ helped in accelerating the NR curing reaction. Microscopical analysis revealed a homogenous dispersion of fillers in the NR matrix. The resultant NR/carbon black/hyperbranched polyester-grafted Al$_2$O$_3$ nanocomposite based tread compound exhibited enhanced thermal stability as well as mechanical properties such as tensile strength and modulus, breaking elongation, tear strength, hardness and abrasion resistance. Yang *et al.* [54] investigated the usefulness of graphene and GO to improve the performance of carbon black filled NR with an aim to develop engineering materials. The introduction of graphene as well as GO was beneficial to improve carbon black dispersion in the NR matrix. Both dynamic and static mechanical properties of carbon black filled NR were observed to significantly increase after the incorporation of small amount of graphene and GO, whereas higher concentration had a detrimental effect due to nanofiller flocculation. Using graphene, the heat build-up and flex cracking tendencies of CB filled NR were diminished with increasing concentration.

Sarkawi *et al.* [55] investigated the effect of graphene on the performance of nano-silica and carbon black reinforced NR and ENR composites for tire applications. Graphene incorporation significantly influenced the performance of nano-silica filled ENR as a result of the combined effect of nano-silica dispersion, interactions between ENR and filler and augmented bound rubber amount. A high wet skid resistance was also observed for the nanocomposite. In another study, Lu *et al.* [56] reported MWCNTs reinforced NR composite. After silanization, MWCNTs were dispersed homogeneously in the NR matrix with improved polymer-filler interactions. NR/MWCNTs nanocomposites exhibited superior mechanical characteristics, thermally conductive nature and reduced volume resistivity. The automotive tire designed using the nanocomposite displayed high fuel efficiency, fatigue strength, improved crack-growth resistance and excellent anti-static behavior.

Ismail *et al.* [57] incorporated MWCNTs to achieve silica dispersion in the NR matrix for tire applications. On incorporating MWCNTs, the curing and scorch time of the resultant nanocomposites were reduced. In addition, the tensile properties and fatigue resistance of the resultant nanocomposites were increased with increase in MWCNTs content up to 1 phr due to the uniform filler dispersion. The highest tensile strength was observed at a loading ratio of 29 phr silica/1 phr MWCNTs. With a goal to enrich the performance of automobile tire tread materials based on NR/PB blend, Poikelispaa *et al.* [58] reported the substitution of a portion of carbon black with MWCNTs. In this system, the substitution of up to 5 phr of carbon black with MWCNTs exhibited homogeneously dispersed fillers in the elastomer blend matrix, which significantly enhanced the mechanical performance, attributed to the strong interfacial interactions. The electrical conductivity was also observed to be augmented with the addition of MWCNTs. Ageing behavior of the nanocomposite vulcanizate revealed the generation of additional crosslinks in the network.

3.2.3 Butyl Rubber Nanocomposites

Butyl rubber (BR) is commonly used to make tire inner liner and tire inner tubes, due to its outstanding impermeability. Nanomaterial incorporation is commonly employed to enrich BR performance for tire applications. In one such study, Kumar *et al.* [59] reported the potential of quaternary ammonium-treated MMT reinforced brominated BR (BBR) based compound for developing automotive tubeless tire inner liner. Functionalized MMT significantly enhanced the performance including barrier, dynamic and static mechanical properties. The nanocomposite based tire inner liner compound displayed lower hysteresis and running temperature together with improved fuel efficiency. In another study, the authors investigated the suitability of BBR/epichlorohydrin rubber/carbon black/organo-MMT nanocomposite for aforementioned application. Due to the synergistic effect of hybrid filler microstructure, homogeneous filler distribution and MMT exfoliation, the resultant nanocomposite system displayed enhanced modulus (up to 54%), tear strength (up to 20%), hydrophobicity and a remarkable decrease in rate of water vapor transmission (up to 20% reduction) [60].

Halim *et al.* [61] reported the addition of nano-clay to modify the mechanical properties of BR for automotive tire applications as well as to achieve effectiveness in terms of cost, service, time, and energy. The use of hexamine : resorcinol : hydrated silica (HRM) bonding system was favorable to achieve homogeneous organo-MMT dispersion in the BR matrix, which correspondingly resulted in enhanced mechanical and curing properties, as compared to HRM-free system.

He *et al.* [62] studied the effect of GO distribution on the barrier and dielectric properties of BBR/GO nanocomposites for generating effective tire inner liner. Compared to BBR nanocomposite with uniformly dispersed GO possessing non-segregated morphology, BBR nanocomposite with spatially distributed GO possessing segregated morphology demonstrated higher gas impermeability and lower electrical percolation threshold even with a GO fraction of less than 0.005, owing to increased equivalent aspect ratio.

Xiong *et al.* [63] reported the incorporation of ionic liquid functionalized GO into BBR matrix to generate superior thermal stability for tire inner liner application. The thermal properties such as degradation temperature and glass transition temperature of BBR/ionic liquid-functionalized GO nanocomposites were increased as a result of enhanced GO exfoliation and resultant interfacial interactions. In addition, the thermal conductivity of the nanocomposite with 4 wt% GO-IL was observed to be 1.3 times higher than the unfilled polymer.

For developing high performance automotive tire inner liner, Kotal *et al.* [64] discussed the usefulness of bromobutyl rubber (BIIR) - grafted functionalized graphene reinforced BIIR nanocomposites. Owing to combined effect of enhanced exfoliation of fillers and interfacial interactions, significant enhancement in tensile strength (200%), storage modulus (189%), thermal stability (17 °C), permittivity (460%) and gas impermeability (44%) were observed as compared to neat BIIR.

In order to strengthen the tubeless tire inner liner applications of chlorinated isobutyl isoprene rubber (CIIR), Frasca *et al.* [65] incorporated multi-layer graphene (MLG). Addition of only 3 phr of MLG to CIIR resulted in an enhancement in Young's modulus by more than two times, along with reduction in the permeability of O_2 and CO_2 by 30%. In addition, the nanocomposites exhibited enhanced flame resistance, along with rheological and curing properties. In another study, the authors discussed the usefulness of adding multi-layer graphene along with carbon black into CIIR [66]. The addition of 3 phr MLG and 20 phr CB increased the hardness to the same extent as 40 phr CB, as demonstrated in Figure 3.1. The resultant ternary nanocomposite also displayed significant enhancement in weathering resistance.

3.2.4 Other Elastomer Nanocomposites

Kumar *et al.* [67] reported the modification of synthetic isoprene rubber by incorporating graphene. The presence of graphene in the nanocomposite vulcanizate composition significantly improved mechanical, rheological and electrical properties. A reduction in scorch time with increase in graphene content was also observed in the nanocomposites.

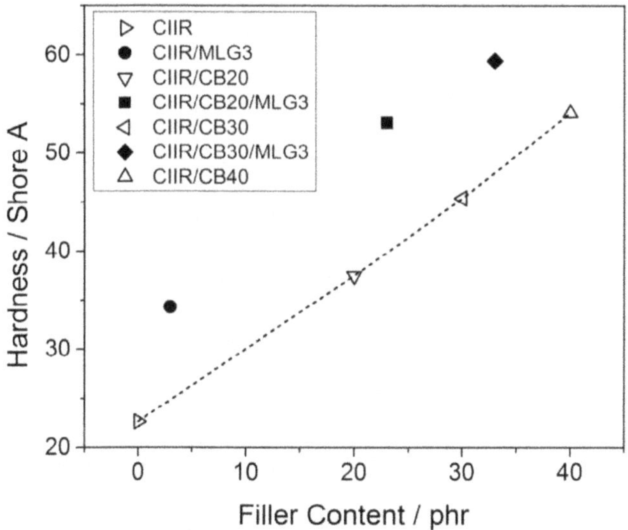

Figure 3.1 Hardness of CIIR and its composites as a function of the filler content. Reproduced from Reference 66 with permission from MDPI.

Zhang *et al.* [68] discussed the usefulness of octadecylamine-functionalized GO to widen the automotive tire applicability of PB. The toughness and elongation of the nanocomposites increased by 332% and 191% respectively as compared to pure polymer. However, the enhancement was at the expense of tensile modulus due to the crumpling effect of GO sheets. The use of octadecylamine-functionalized GO was also observed to considerably enhance the oxidation resistance of PB.

Lei *et al.* [69] described the development of energy effective automotive tire material based on silica nanoparticles reinforced poly(di-*n*-butyl itaconate-*co*-butadiene). The sustainable polymer nanocomposite tire compound displayed superior anti-wear behavior, low resistance towards rolling and outstanding wet traction.

In another study, Qiao *et al.* [70] revealed that the potential of nano-silica reinforced epoxidized bio-based poly(dibutyl itaconate-*ter*-isoprene-*ter*-glycidyl methacrylate) for tire applications. The authors observed effective silica dispersion in the epoxidized polymer matrix due to the ring opening reaction (Figure 3.2). Improved mechanical properties were observed for the nanocomposites in both static and dynamic modes compared to unmodified version. With the same silica content, the inclusion of 3.7 wt% glycidyl methacrylate (GMA) was also observed to increase the modulus at 100% strain by 150%, whereas the modulus increased by 152.3% at 300% strain.

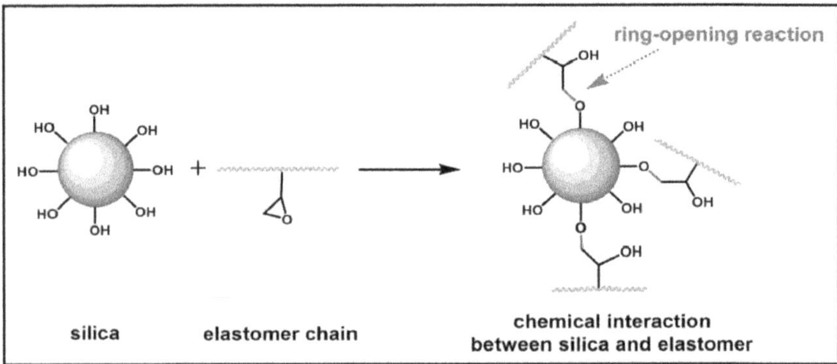

Figure 3.2 Schematic demonstration of covalent linkage between silica and epoxidized elastomer in the nanocomposite system. Reproduced from Reference 70 with permission from American Chemical Society.

3.3 Polymer Nanocomposites for Automotive Tire Inner Tubes

The use of nanomaterials in the formulation of automotive tire inner tubes is beneficial to enhance impermeability and mechanical properties. In a related study, Sadasivuni *et al.* [71] compared the effect of graphene and nano-clay on the gas permeability and mechanical properties of poly(isobutylene-co-isoprene) (IIR) blended with maleic anhydride grafted poly(isobutylene-co-isoprene) (MA-g-IIR). The highest moduli, tensile strength and elongation at break were observed for 60:40 ratio of MA-g-IIR (grafting degree 0.75)/IIR mixture with 5 phr of cloisite 10A.

In another study, Razzaghi-Kashani *et al.* [72] employed organo-MMT to enhance the performance of BR vulcanizate based tire inner tubes. The organo-MMT was beneficial in improving the mechanical properties and gas impermeability at a filler concentration of 3 phr owing to optimal dispersion and intercalation.

Rajasekar and Das [73] reported the use of chlorobutyl rubber (CBR) as an effective compatibilizer to achieve nano-clay exfoliation in BR. The resultant ternary nanocomposites exhibited higher static and dynamic mechanical properties and gas impermeability compared to uncompatibilized version. Overall, CBR compatibilized BR/clay nanocomposites displayed potential for use as a high performance automotive tire inner tube materials. In a similar study, Saritha *et al.* [74] revealed the potential of organo-clay reinforced CBR for the fabrication of tire inner tubes. The mechanical properties of CBR improved with the addition of nano-clay up to 10 phr.

Azizli *et al.* [75] reported the development of BR/chloroprene rubber/organo-MMT nanocomposites. Enhancement in mechanical properties like tensile strength (100%), modulus (300%), elongation at break (200%) and resilience with increase in organo-MMT and chloroprene rubber content were observed, owing to combined effect of filler dispersion and polymer intercalation into the layered structure of organo-MMT. The presence of organo-MMT had also a negative influence on the curing behavior of nanocomposites, as longer scorch and cure durations were observed. In another study, the authors observed similar behavior for BR/EPDM/organo-MMT nanocomposites [76].

Meera *et al.* [77] examined organo-MMT/NR nanocomposites for effective gas barrier properties for tire inner tube applications. Addition of organo-MMT was advantageous in enhancing the NR impermeability towards various gases such as oxygen, carbon dioxide and nitrogen due to the generation of tortuous path resulting from improved interactions of organo-MMT with NR and reduced free volume. Yaragalla *et al.* [78] reported graphene and graphite reinforced ENR compounds for high performance automotive tire inner tubes. The use of both graphene and graphite was helpful in improving the barrier and dielectric features of ENR. The extent of improvement was observed to be higher for ENR nanocomposites with graphene, attributed to the higher degree of hydrogen bonding. For instance, the nanocomposite containing ENR with 50% epoxidation degree and 2 wt% graphene demonstrated optimal oxygen impermeability.

3.4 Polymer Nanocomposites for Automotive Coatings

Addition of nanomaterials in conventional automotive coating formulations is beneficial to optimize their performance, especially scratch resistance, weather resistance, impermeability and durability.

3.4.1 Acrylic/Melamine Nanocomposite Coatings

Yari *et al.* [79] reported high performance automotive clear coat based on hydroxylated nano-polyhedral oligomeric silsesquioxane (nano-POSS) reinforced acrylic/melamine nanocomposites. Due to the improved crosslink density and hardness, a significantly enhanced scratch resistance was observed for the nanocomposite clear coat. In addition, the incorporation of nano-POSS improved the healing capability due to the formation of interfacial hydrogen bonding. Interestingly, the optical quality of the acrylic/melamine clear coat remained unaltered after the inclusion of nano-POSS. In another study, the authors revealed the ability of nano-POSS to enhance the toughness of acrylic/melamine thermoset automotive clear coat without affecting other properties [80]. Also, the use of hydroxylated nano-POSS/hydroxylated hyperbranched polymer hybrid was observed to be useful in enhancing both healing potential and resistance to scratching of acrylic/melamine automotive clear coat concurrently due to the presence of both physical and covalent crosslinks [81].

In a similar study, Yari *et al.* [82] reported the potential of nanosilica in improving the weatherability of acrylic/melamine automotive clear coat. In another study, Ramezanzadeh *et al.* [83] obtained higher scratch resistance for nano-SiO_2 reinforced acrylic/melamine automotive clear coat, as compared to neat clear coat. The authors also examined the scratch morphology of the nanocomposite automotive clear coat and observed fracture and plastic deformations at low and high amount of filler incorporation respectively. Tahmassebi *et al.* [84] also reported similar scratch behavior for acrylic/melamine/silica nanocomposite clear coat.

3.4.2 Polyurethane Nanocomposite Coatings

Mirabedini *et al.* [85] studied the weathering behavior of silanized nano-TiO_2 filled polyurethane (PU) coat. The weatherability of PU was observed to improve after the addition of 0.5 and 1 wt% silanized nano-TiO_2 due to the diminished photocatalytic activity. Silane treatment of nano-TiO_2 particles improved their dispersion in the PU matrix, which correspondingly resulted in enhanced mechanical properties. In another study, Hang *et al.* [86] reported the usefulness of nano-TiO$_2$ to enhance the UV resistance of PU coatings. Higher enhancement in UV resistance for PU nanocomposite coating with 0.1 wt% of silanized nano-TiO$_2$ was observed as compared to the nanocomposite coating with unmodified nano-TiO$_2$.

For optimizing the properties of PU coatings, Palimi *et al.* [87] reported the use of silanized nano-Fe$_2$O$_3$ particles. Silanization of Fe$_2$O$_3$ was useful in improving its dispersion in the PU coating. Enhanced toughness, crosslink density and T_g were noted for silanized Fe$_2$O$_3$/PU nanocomposite coating and the composite containing 3 g of silane per 5 g of nano-Fe$_2$O$_3$ presented optimal enhancement. Lahijania *et al.* [88] compared the effect of untreated and silanized silica nanoparticles on the performance of PU acrylate coatings. The optimal bonding of silanized nano-silica in the PU acrylate matrix revealed its strong reinforcing ability, as compared to untreated version. As a result, the properties such as hardness, elastic recovery, abrasion resistance and T_g of the silanized nano-silica/PU acrylate nanocomposite coating were observed to enhance.

Yahyaei and Mohseni [89] reported the generation of silica/PU acrylate nanocomposite hard coatings on the surface of polycarbonate using sol-gel process for achieving effective automotive coating application. Owing to higher inorganic concentration and lower silane acrylate concentration, the resultant nanocomposite coating exhibited higher elastic modulus, hardness, scratch resistance and abrasion resistance in comparison to neat PU acrylate coating. Verma *et al.* [90] detailed the generation of PU/organo-clay nanocomposite coatings in order to achieve scratch resistant coating application. An improvement of 22% in scratch resistance was observed with 5 wt% organo-clay in the PU. No notable change in the color of the PU coating was observed up to 5 wt% of nano-clay. In another research, Gurunathan *et al.* [91] reported enhanced tensile strength and modulus as well as thermal stability for bio-sourced PU/organo-clay nanocomposite coatings.

Ghermezcheshme *et al.* [92] reported the development of POSS/PU nanocomposite coatings. POSS containing higher level of hydroxyl functionalities enhanced the scratch resistance, whereas POSS with lower extent of hydroxyl functionalities had a diminishing effect. Norouzi *et al.* [93] examined the impact of acrylated nano-POSS on the properties of PU acrylate hard coatings. In comparison to nanocomposite coatings with unmodified nano-POSS, the transparency of acrylated nano-POSS reinforced PU acrylate nanocomposite coatings was superior. In addition, a superior abrasion resistance for PU acrylate/acrylated nano-POSS nanocomposite coatings was observed owing high surface hardness. However, the micro hardness and storage modulus of PU acrylate coatings were observed to diminish after the incorporation of acrylated nano-POSS, resulting from enhanced free volume.

Nuraje *et al.* [94] prepared PU/graphene top coatings with varying graphene content. The coating with 2% graphene was observed to optimally enhance the UV stability and corrosion resistance of PU top coat. Liao *et al.* [95] highlighted low percolation for 0.15 wt% graphene reinforced PU acrylate nanocomposite coatings. Graphene induced reinforcement in the rubbery region of PU acrylate coatings was observed, whereas the reinforcement in the glassy region was not significant. In another study, Miraftab *et al.* [96] employed graphene/azo dye hybrid in the formulation of PU coatings to achieve superior UV resistance.

3.4.3 Acrylic Nanocomposite Coatings

Dong and Liu [97] studied the effect of silanized GO on the performance of acrylic coatings. Owing to optimal silanized GO dispersion in the acrylic coatings and resultant interfacial bonding, enhanced mechanical and thermal properties as well as chemical resistance were obtained. Al-Kawaz *et al.* [98] employed acrylated MWCNTs to effectively reinforce acrylic coatings. Improved mechanical and tribological performance were observed for the developed acrylated MWCNTs/acrylic nanocomposite coatings as a result of uniform filler dispersion.

Kugler *et al.* [99] compared the usefulness of graphene, CNTs and their blend in modifying the properties of acrylic coatings. The addition of carbonaceous filler systems enhanced the hardness, modulus, thermal degradation temperature and electrical conductivity of the acrylic coatings without affecting adhesion strength and T_g. The extent of enhancement in hardness and modulus was higher for CNTs/graphene hybrid reinforced acrylic coatings, whereas thermal stability and electrical conductivity were higher for CNTs reinforced system. In another study, the authors investigated the effect of different inorganic nanofillers into CNTs reinforced acrylic coatings [100]. The enhancement in surface conductivity and transparency was higher for silica/CNTs/acrylic nanocomposite coatings, as compared to alumina and titania filled CNTs/acrylic nanocomposite coatings and neat CNTs/acrylic nanocomposite coating. Among various nanocomposite coatings, silica and titania filled CNTs/acrylic nanocomposite coatings exhibited higher hardness improvement and gloss reduction. In addition, silica and alumina filled CNTs/acrylic nanocomposite coatings had higher thermal stability.

Motlagh *et al.* [101] studied the effect of silver/TiO$_2$ hybrid nanoparticles on the properties of PU acrylate clear coatings. The hardness and abrasion resistance of the nanocomposites displayed an increasing tendency only at low filler loading, whereas the scratch resistance increased continuously with increase in the filler loading. Also, superior antibacterial behavior was noted for the nanocomposite coatings. Chouwatat *et al.* [102] reported nano-POSS reinforced acrylic nanocomposites for functional automobile hard coating applications. Superior scratch resistance was observed for the nanocomposite hard coatings with the amount of nano-POSS lower than 10 wt%. However, the acrylic nanocomposites containing higher amount of nano-POSS exhibited poor scratch resistance attributed to the incomplete crosslinking reaction caused by substantial steric hindrance.

3.4.4 Polyester Nanocomposite Coatings

Lee *et al.* [103] reported the usefulness of polyester/organo-MMT nanocomposite coatings for the effective passivation of pre-coated automotive metal parts. Microscopical and X-ray diffraction analyses revealed partial intercalation and exfoliation of organo-MMT in the polymer coatings. Mechanical properties such as stiffness, tensile strength and formability of the coatings were observed to increase with increase in organo-MMT. T_g of the cured nanocomposite coatings was reduced as well. Salt spray test result of the polyester/organo-MMT nanocomposites revealed improved resistance to corrosion owing to reduced permeability resulting from the longer diffusion path.

In another study, Konwar *et al.* [104] reported acrylate-grafted vegetable oil sourced polyester/organo-MMT nanocomposite coatings. The thermal degradation temperature and tensile strength of acrylate-grafted bio-sourced polyester were observed to significantly increase after the addition of organo-clay. Also, the rheological analysis revealed shear thinning behavior for the nanocomposite coatings. Improved biodegradability was also observed for acrylate-grafted bio-sourced polyester after organo-clay incorporation.

3.5 Polymer Nanocomposites for Automotive Polymer Electrolyte Membrane Fuel Cells

To develop energy conversion systems possessing better consistency and sustainability for automotive applications, polymer matrix nanocomposites have displayed significant suitability owing to desirable power/energy performance, lightweight structure and cost effectiveness. In the light of effective automotive applications, polymeric electrolyte membrane fuel cells (PEMFCs) represent a foremost category of environmentally amiable fuel cells receiving wide acceptance [105,106].

To obtain enhanced proton conductivity at moderate range of temperature and relative humidity, Farrukh *et al.* [107] reported the generation of Nafion/poly(monomethoxy oligo(ethylene glycol) methacrylate)-modified silica nanocomposite membranes. A superior proton conductivity for Nafion nanocomposite containing 1 wt% of modified silica nanoparticles was observed. Jia *et al.* [108] reported the performance optimization of Nafion proton exchange membranes by generating Nafion blended Nafion/boron nitride nanocomposite (NBN). Due to the formation of continuous long-range nanochannels, the resultant nanocomposite membrane demonstrated about 6 fold improvement in proton conductivity at high temperature and sub-ambient humid conditions (Figure 3.3).

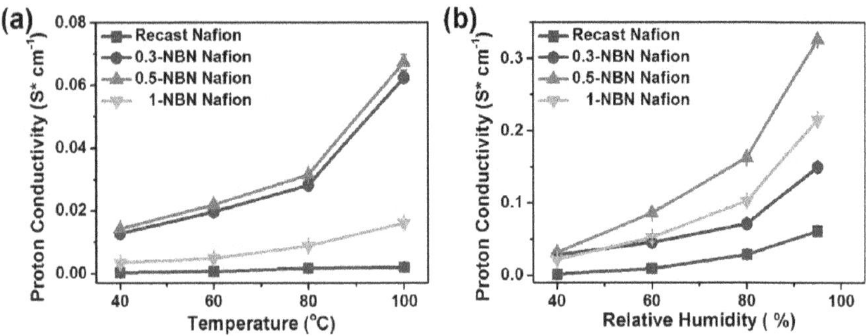

Figure 3.3 Proton conductivities of Nafion/Nafion-BN nanocomposite proton exchange membranes: (a) temperature-dependent at 40% RH, and (b) humidity-dependent at 80°C. Reproduced from Reference 108 with permission from American Chemical Society.

Zhang *et al.* [109] dispersed phosphonic acid-modified graphene into Nafion to generate effective proton conducting membranes for

PEMFCs. The resultant acidified graphene/Nafion nanocomposite membranes exhibited significantly increased proton conductivity under high temperature and ambient humid conditions. For instance, the nanohybrid membrane containing 2 wt% modified graphene exhibited a proton conductivity of 0.277 S cm^{-1} at 100 °C and 100% RH, and 0.0441 S cm^{-1} at 80 °C and 40% RH, which were observed to be 1.2 and 6.6 times higher than pristine Nafion membrane. In addition, superior power density as well as open-circuit voltage were observed for the nanocomposite membranes.

Peng *et al.* [110] applied atom transfer radical addition reaction to generate strong interaction of GO with Nafion for PEMFC applications. Improved proton conductivity (1.6 folds) and fuel cell performance (35-40%) for the developed Nafion/graphene nanocomposite membrane were observed. Kowsari *et al.* [111] introduced ionic liquid modified GO with phosphonic acid doping into sulfonated polyimide (sPI) with an aim to yield high performance proton conducting membranes. The resultant sPI/ionic liquid-modified GO (5%) nanocomposite membrane exhibited maximum proton conductivity values of 0.0772 S cm^{-1} (at 160 °C and ambient humidity) and 0.1243 S cm^{-1} (at 120 °C and 80% RH). In another study, He *et al.* [112] investigated the effect of GO size on the performance of fuel cell containing GO/sPI nanocomposite membranes and observed maximal property enhancement for the composite containing smallest sized GO (0.5% fraction; high conductivity (1.2 S cm^{-1} at 80 °C and RH 100%), low methanol permeability (1.07×10^{-7} cm^2 S^{-1} at 25 °C) and outstanding fuel cell performance).

Zhao *et al.* [113] revealed the potential of quaternized GO reinforced sulfonated poly(ether sulfone) (sPES) nanocomposite membranes for effective PEMFC applications. The composite with 10% modified filler exhibited high proton conductivity of about 0.08 S cm^{-1} at 80 °C. The composite membranes also exhibited enhanced mechanical properties, oxidation resistance and methanol impermeability. Quaternized GO was also beneficial to reduce water absorption and swelling ratio in the nanocomposite membranes.

Gahlot *et al.* [114] detailed the development of proton conducting membranes with the help of sulfonic acid-treated GO reinforced sPES nanocomposites. A homogeneous sulfonated GO dispersion within the sPES matrix was observed, which resulted in improved interfacial interactions. Significantly enhanced proton conductivity, ion-exchange capacity, methanol impermeability and water retention was noted for the nanocomposite membranes (Figure 3.4).

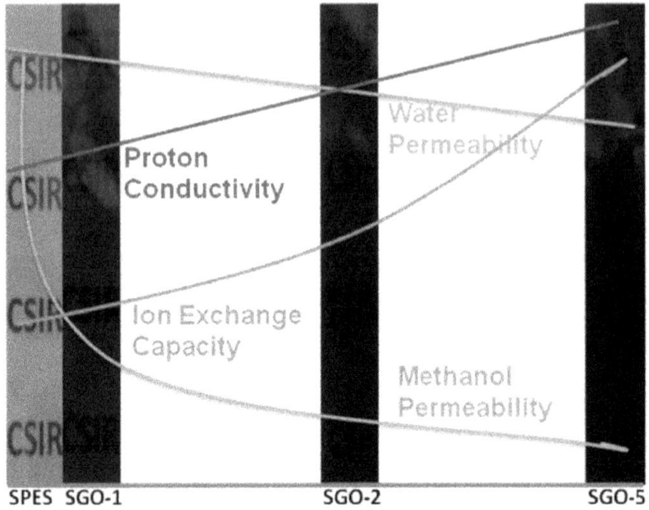

Figure 3.4 Performance of sulfonated poly(ether sulfone)/sulfonated GO (SGO) nanocomposite proton exchange membrane as a function of SGO content (0, 1, 2, and 5 wt%). Reproduced from Reference 114 with permission from American Chemical Society.

Miao *et al.* [115] reported the generation of high performance proton conducting membranes for PEMFC applications using sulfonic acid-treated GO reinforced sulfonated poly(arylene ether sulfone) (sPAES) nanocomposites. The mechanical and fuel cell properties of 5 wt% sulfonic acid-treated GO incorporated sPAES nanocomposite were superior in comparison to Nafion. In another study, Feng *et al.* [116] fabricated high performance proton conducting membranes using GO/CNTs hybrid reinforced sulfonated poly(arylene ether nitrile) (sPAEN). In comparison to individual GO and CNTs, the use of GO/CNTs hybrid filler system was advantageous in avoiding the self-flocculation of either GO or CNTs in the sPAEN matrix. A uniform dispersion of GO/CNTs hybrid in the sPAEN matrix was observed, thus, resulting in the systematic generation of proton transport pathways. The GO/CNTs/SPEN nanocomposite membrane with the ratio of 2:2 exhibited the highest proton conductivity of 0.1197 S/cm at 20 °C. In addition, high methanol impermeability of sPAEN was observed to remain unaltered after GO/CNTs addition.

Oh *et al.* [117] reported that the addition of boron nitride nanoflakes (BNNFs) even at a low content (0.3 wt %) significantly enhanced the mechanical stability of sulfonated poly(ether ether ketone) (sPEEK). BNNFs were also non-covalently functionalized with pyrenesulfonic acid (PSA) to enhance the performance. Figure 3.5 depicts the synthesis procedure for the composite membranes.

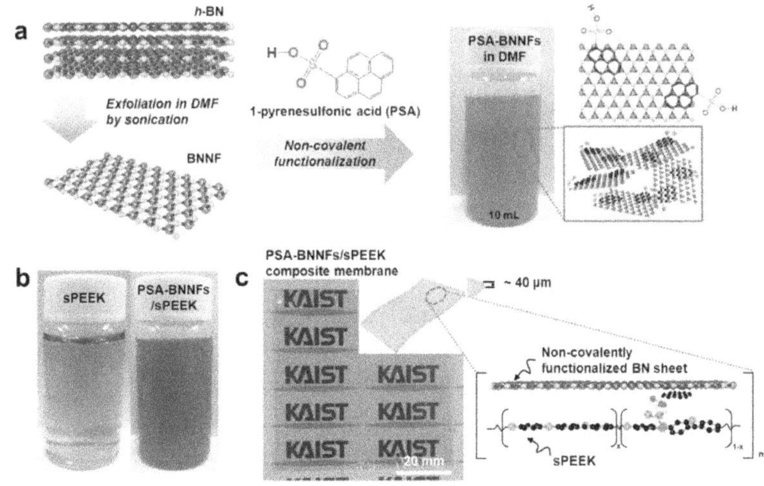

Figure 3.5 Schematic diagram depicting the overall procedure for the preparation of PSA-BNNFs and PSA-BNNFs/sPEEK composite membrane. Reproduced from Reference 117 with permission from American Chemical Society.

Beydaghi *et al.* [118] reported the potential of sulfonic acid-treated GO reinforced poly(vinyl alcohol) (PVA) crosslinked nanocomposite proton exchange membranes. For a sulfonated GO loading of 5 wt%, the resultant PVA nanocomposite exhibited substantially enhanced proton conductivity (0.050 S cm^{-1}), tensile strength (67.8 MPa) and thermal stability (melting temperature, T_m = 223 °C). In addition, a superior power density of 16.15 mW cm^{-2} at 30 °C for PEMFC based on the developed nanocomposite was observed.

Liu *et al.* [119] fabricated sulfonated GO/chitosan nanocomposite based proton conductive membranes for effective PEMFC application. Owing to efficient dispersion of sulfonated GO within the chitosan matrix, the resultant nanocomposite membranes displayed enhanced mechanical and thermal properties. Additionally, the nanocomposite membrane with 2.0% sulfonated GO exhibited a 122.5% increase in hydrated conductivity and a 90.7% increase in anhydrous conductivity. In another study, Shirdast *et al.* [120] used sulfonated GO/sulfonated chitosan hybrid to optimize the proton exchange property of chitosan. The proton conductivity and proton selectivity of chitosan displayed about 6 times enhancement after the addition of 5 wt% sulfonated GO/10 wt% sulfonated chitosan hybrid. The thermal and mechanical stabilities of the nanocomposite membrane were enhanced as well. Bai *et al.* [121] incorporated phosphorylated graphene oxide (PGO) nanosheets into chitosan matrix to achieve nanohybrid membranes. The nanohybrid membrane with 2.5% PGO content exhibited a 22.2-time increase in conductivity from 0.25 mS cm^{-1} to 5.79 mS cm^{-1} (160 °C, 0% RH).

3.6 Polypropylene (PP) Nanocomposites for Automotive Applications

PP is a versatile thermoplastic polymer which has been widely used to generate various automotive parts such as bumpers, dashboard, wheel housing, battery case, guide channels, side panels, crash panel, under-hood parts, containers, body panel, lighting, air filter housing, interior trim, exterior trim, seats, etc. [3]. A number of studies have reported further enhancement of the performance of PP for the aforementioned automotive applications by incorporating nanomaterials. Nanomaterial incorporation has been reported to be beneficial in strengthening toughness/stiffness balance, thermal stability, charge dissipation, barrier properties and weight saving of PP [23]. For developing high performance PP nanocomposites, graphene and its different analogs have found immense acceptance recently, and some examples of these PP nanocomposite systems include PP/GO [122,123], PP/graphene nanosheets [124], PP/graphene nanoplatelets [125-129], PP/long-chain alkylamine-functionalized GO [130], PP/hexamethylene diamine-modified GO [131], PP/covalently-modified graphene [132], PP/low molecular weight PP-functionalized graphene [133], PP/styrene-(ethylene-*co*-butylene)-styrene (SEBS)/graphene nanoplatelets [134], PP/polystyrene/SEBS/graphene nanoplatelets [135], PP/kenaf flour/graphene nanoplatelets [136], PP/CNTs/graphene nanoplatelets [137] and PP/silica/graphene nanoplatelets [138].

3.7 Poly(lactic acid) (PLA) Nanocomposites for Automotive Applications

PLA is an important class of bio-based aliphatic polyesters, which has the potential to replace petroleum sourced polymers owing to its superior physico-mechanical features, minimal carbon footprint, availability, inexpensiveness and recyclability. PLA has received considerable research attention for numerous automotive applications through nanomaterial incorporation, which is advantageous to optimize the performance in terms of stiffness, toughness, thermal properties and crystallinity [139].

Notta-Cuvier *et al.* [140] reported the fabrication of halloysite nanotubes (HNTs) reinforced PLA nanocomposites in the presence of tributyl citrate (TBC) plasticizer for effective automobile applications. The authors observed comprehensive mechanical properties including impact strength, tensile strength, flexural strength, rigidity, ductility and crystallinity for PLA nanocomposite containing 9 wt% HNTs and 10 wt% TBC. The study confirmed the potential of PLA/HNTs/TBC nanocomposites for superior automotive applications as a sustainable substitute to conventional petroleum based polymers. In another study, the authors prepared PLA/TBC/ethylene copolymer Biomax Strong®/organo-clay nanocomposites for achieving automotive applications. Improved tensile strength and rigidity as well as ductility were observed for the developed PLA nanocomposites.

Guo *et al.* [142] reported the generation of alkali-modified HNTs reinforced PLA nanocomposites with an aim to achieve effective automobile applications. Improved crystallinity and reduced crystallization temperature were observed for the PLA nanocomposites. Alkali modification of HNTs was useful to achieve optimal dispersion within the PLA matrix. As a result, improvements in thermal stability, tensile strength and tensile modulus for alkali-modified HNTs/PLA nanocomposites were observed in comparison to nanocomposites with unmodified HNTs. In another study, Murariu *et al.* [143] used silanized HNTs to determine the automotive practicability of PLA. Silanization of HNTs helped to achieve effective dispersion in the PLA matrix even at higher loading. Enhanced static and dynamic mechanical properties were observed for the PLA nanocomposites. Prashantha *et al.* [144] reported a higher extent of enhancement in tensile strength, impact strength, flexural strength and rigidity without any marked change in ductility for quaternary ammonium salt-modified HNTs reinforced PLA nanocomposites compared to neat HNTs/PLA nanocomposites.

Tham *et al.* [145] developed PLA/maleic anhydride-grafted-SEBS (SEBS-*g*-MA)/HNTs nanocomposites and observed significant improvement in impact strength and thermal stability after the addition of SEBS-*g*-MA into PLA/HNTs nanocomposite. In another study, the authors investigated the effect of *N,N'*-ethylenebis(stearamide) (EBS) addition on the performance of PLA/HNTs nanocomposites. The use of EBS improved the dispersion of halloysite within the PLA matrix, which correspondingly resulted in enhanced crystallinity, thermal stability, impact resistance and breaking extension. PLA nanocomposite containing 2 wt% of HNTs and 5 wt% of EBS exhibited the overall best performance.

3.8 Conclusions

Polymer nanocomposites are extensively used for various automotive applications to provide functional features such as weight reduction, fuel efficiency, performance, safety and low exhaust emissions owing to their attractive properties such as superior strength to weight ratio, ease of processing, low permeability, etc. In this chapter, recent advancements in automotive applications of various polymer nanocomposites have been reviewed in detail. Overall, the use of polymer nanocomposites in automotive tires, automotive tire inner tubes, automotive coatings and polymer electrolyte membrane fuel cells has been observed to be beneficial in improving their performance in many ways. In addition, the extent of improvement is greatly influenced by many factors like nature of polymers and nanomaterials, dispersion state of nanomaterials within the polymer matrix, interfacial interactions, etc.

References

1. Naffakh, M., Diez-Pascual, A. M., Marco, C., Ellis, G. J., and Gomez-Fatou, M. A. (2013) Opportunities and challenges in the use of inorganic fullerene-like nanoparticles to produce advanced polymer nanocomposites. *Progress in Polymer Science*, **38**, 1163-1231.
2. Zaman, I., Manshoor, B., Khalid, A., and Araby, S. (2014) From clay to graphene for polymer nanocomposites - A survey. *Journal of Polymer Research*, **21**, 429.
3. Lyu, M.-Y., and Choi, T. G. (2015) Research trends in polymer materials for use in lightweight vehicles. *International Journal of Precision Engineering and Manufacturing*, **16**, 213-220.
4. Garces, J. M., Moll, D. J., Bicerano, J., Fibiger, R., and McLeod, D. G. (2000) Polymeric nanocomposites for automotive applications. *Advanced Materials*, **12**, 1835-1839.

5. Zhang, S., Wang, H. L., Ong, S. E., Sun, D., and Bui, X. L. (2007) Hard yet tough nanocomposite coatings–present status and future trends. *Plasma Processes and Polymers*, **4**, 219-228.

6. Coelho, M. C., Torrao, G., and Emami, N. (2012) Nanotechnology in automotive industry: research strategy and trends for the future-small objects, big impacts. *Journal of Nanoscience and Nanotechnology*, **12**, 6621-6630.

7. Paul, D., and Robeson, L. M. (2008) Polymer nanotechnology: Nanocomposites. *Polymer*, **49**, 3187-3204.

8. Kotal, M., and Bhowmick, A. K. (2015) Polymer nanocomposites from modified clays: Recent advances and challenges. *Progress in Polymer Science*, **51**, 127-187.

9. Fan, J., Shi, Z., Zhang, L., Wang, J., and Yin, J. (2012) Aramid nanofiber-functionalized graphene nanosheets for polymer reinforcement. *Nanoscale*, **4**, 7046-7055.

10. Varghese, A. M., and Mittal, V. (2017) Nanotubes. In: *Kirk-Othmer Encyclopedia of Chemical Technology*, John Wiley & Sons, USA, doi:10.1002/0471238961.koe00026.

11. Lim, Y., Tan, Y., Lim, H., Huang, N., and Tan, W. (2013) Preparation and characterization of polypyrrole/graphene nanocomposite films and their electrochemical performance. *Journal of Polymer Research*, **20**, 156.

12. Azeez, A. A., Rhee, K. Y., Park, S. J., and Hui, D. (2013) Epoxy clay nanocomposites–processing, properties and applications: A review. *Composites, Part B: Engineering*, **45**, 308-320.

13. Presting, H., and Konig, U. (2003) Future nanotechnology developments for automotive applications. *Materials Science and Engineering C*, **23**, 737-741.

14. Pavlidou, S., and Papaspyrides, C. (2008) A review on polymer–layered silicate nanocomposites. *Progress in Polymer Science*, **33**, 1119-1198.

15. Goettler, L., Lee, K., and Thakkar, H. (2007) Layered silicate reinforced polymer nanocomposites: development and applications. *Polymer Reviews*, **47**, 291-317.

16. Gul, S., Kausar, A., Muhammad, B., and Jabeen, S. (2016) Research progress on properties and applications of polymer/clay nanocomposite. *Polymer-Plastics Technology and Engineering*, **55**, 684-703.

17. Kojima, Y., Usuki, A., Kawasumi, M., Okada, A., Kurauchi, T., Kamigaito, O., and Kaji, K. (1994) Fine structure of nylon 6-clay hybrid. *Journal of Polymer Science, Part B: Polymer Physics*, **32**, 625-630.

18. Okada, A., and Usuki, A. (2006) Twenty years of polymer-clay nanocomposites. *Macromolecular Materials and Engineering*, **291**, 1449-1476.

19. Zeng, Q., Yu, A., Lu, G., and Paul, D. (2005) Clay-based polymer nan-

ocomposites: research and commercial development. *Journal of Nanoscience and Nanotechnology*, **5**, 1574-1592.

20. Muller, K., Bugnicourt, E., Latorre, M., Jorda, M., Echegoyen Sanz, Y., Lagaron, J.M., Miesbauer, O., Bianchin, A., Hankin, S., Bolz, U., Perez, G., Jesdinszki, M., Lindner, M., Scheuerer, Z., Castello, S., and Schmid, M. (2017) Review on the processing and properties of polymer nanocomposites and nanocoatings and their applications in the packaging, automotive and solar energy fields. *Nanomaterials*, **7**, 74.

21. Chandra, A. K., and Bhandari, V. (2013) Nanocomposites for tyre applications. In: *Advances in Elastomers II*, Visakh, P. M., Thomas, S., Chandra, A. K., and Mathew, A. P. (eds.), Springer, Germany, pp. 183-203.

22. Wallner, E., Sarma, D., Myers, B., Shah, S., Ihms, D., Chengalva, S., Parker, R., Eesley, G., and Dykstra, C. (2010) Nanotechnology applications in future automobiles. SAE Technical Paper 2010-01-1149, doi:10.4271/2010-01-1149.

23. Chandra, A. K., and Kumar, N. R. (2017) Polymer nanocomposites for automobile engineering applications. In: *Properties and Applications of Polymer Nanocomposites*, Tripathy, D. B., and Sahoo, B. P. (eds.), Springer, Germany, pp. 139-172.

24. Mohseni, M., Ramezanzadeh, B., Yari, H., and Gudarzi, M. M. (2012) The role of nanotechnology in automotive industries. In: *New Advances in Vehicular Technology and Automotive Engineering*, Carmo, J. (ed.), IntechOpen, UK, doi: 10.5772/49939.

25. Chandra, A. K. (2007) Tire technology: Eecent advances and future trends. *Rubber World*, **236**, 27-32.

26. Li, S., Mao, Y., and Ploehn, H. J. (2017) Mechanical reinforcement in magadiite/styrene-butadiene rubber composites. *Journal of Applied Polymer Science*, **134**, doi: 10.1002/app.44763.

27. Mao, Y., Li, S., and Ploehn, H.J. (2017) The role of interlayer grafting on the mechanical properties of magadiite/styrene-butadiene rubber composites. *Journal of Applied Polymer Science*, **134**, doi: 10.1002/app.45025.

28. Mao, Y., Li, S., Fang, R. L., and Ploehn, H. J. (2017) Magadiite/styrene-butadiene rubber composites for tire tread applications: Effects of varying layer spacing and alternate inorganic fillers. *Journal of Applied Polymer Science*, **134**, 10.1002/app.44764.

29. Kim, W. S., Paik, H. J., Bae, J. W., and Kim, W. (2011) Effect of polyethylene glycol on the properties of styrene-butadiene rubber/organoclay nanocomposites filled with silica and carbon black. *Journal of Applied Polymer Science*, **122**, 1766-1777.

30. Kim, W.-S., Lee, D.-H., Kim, I.-J., Son, M.-J., Kim, W., and Cho, S.-G. (2009) SBR/organoclay nanocomposites for the application on tire tread compounds. *Macromolecular Research*, **17**, 776-784.

31. Song, S. H. (2016) Synergistic effect of clay platelets and carbon nanotubes in styrene-butadiene rubber nanocomposites. *Macromolecular Chemistry and Physics*, **217**, 2617-2625.

32. Tang, Z., Zhang, C., Wei, Q., Weng, P., and Guo, B. (2016) Remarkably improving performance of carbon black-filled rubber composites by incorporating MoS_2 nanoplatelets. *Composites Science and Technology*, **132**, 93-100.

33. Roy, K., Alam, M. N., Mandal, S. K., and Debnath, S. C. (2014) Surface modification of sol-gel derived nano zinc oxide (ZnO) and the study of its effect on the properties of styrene-butadiene rubber (SBR) nanocomposites. *Journal of Nanostructure in Chemistry*, **4**, 133-142.

34. Thaptong, P., Sae-Oui, P., and Sirisinha, C. (2016) Effects of silanization temperature and silica type on properties of silica-filled solution styrene-butadiene rubber (SSBR) for passenger car tire tread compounds. *Journal of Applied Polymer Science*, **133**, doi: 10.1002/app.43342.

35. Qiao, H., Chao, M., Hui, D., Liu, J., Zheng, J., Lei, W., Zhou, X., Wang, R., and Zhang, L. (2017) Enhanced interfacial interaction and excellent performance of silica/epoxy group-functionalized styrene-butadiene rubber (SBR) nanocomposites without any coupling agent. *Composites, Part B: Engineering*, **114**, 356-364.

36. Sun, J., Liu, X., Liang, Y., Wang, L., and Liu, Y. (2017) The preparation of microcrystalline cellulose–$nanoSiO_2$ hybrid materials and their application in tire tread compounds. *Journal of Applied Polymer Science*, **134**, doi: 10.1002/app.44796.

37. Saeed, F., Ansarifar, A., Ellis, R. J., Haile-Meskel, Y., and Irfan, M. S. (2012) Two advanced styrene-butadiene/polybutadiene rubber blends filled with a silanized silica nanofiller for potential use in passenger car tire tread compound. *Journal of Applied Polymer Science*, **123**, 1518-1529.

38. Ostad-Movahed, S., Ansarifar, A., and Song, M. (2009) Effects of different interphases on the mechanical properties of cured silanized silica-filled styrene-butadiene/polybutadiene rubber blends for use in passenger car tire treads. *Journal of Applied Polymer Science*, **113**, 1868-1878.

39. Veiga, V. D.'A., Rossignol, T. M., Crespo, J. d.S., and Carli, L. N. (2017) Tire tread compounds with reduced rolling resistance and improved wet grip. *Journal of Applied Polymer Science*, **134**, doi: 10.1002/app.45334.

40. Malas, A., Pal, P., Giri, S., Mandal, A., and Das, C. K. (2014) Synthesis and characterizations of modified expanded graphite/emulsion styrene butadiene rubber nanocomposites: Mechanical, dynamic mechanical and morphological properties. *Composites, Part B: Engineering*, **58**, 267-274.

41. Yin, B., Zhang, X., Zhang, X., Wang, J., Wen, Y., Jia, H., Ji, Q., and Ding, L. (2017) Ionic liquid functionalized graphene oxide for enhancement of styrene-butadiene rubber nanocomposites. *Polymers for Advanced Technologies*, **28**, 293-302.

42. Yang, Z., Liu, J., Liao, R., Yang, G., Wu, X., Tang, Z., Guo, B., Zhang, L., Ma, Y., and Nie, Q. (2016) Rational design of covalent interfaces for graphene/elastomer nanocomposites. *Composites Science and Technology*, **132**, 68-75.

43. Rooj, S., Das, A., Morozov, I.A., Stockelhuber, K. W., Stocek, R., and Heinrich, G. (2013) Influence of expanded clay on the microstructure and fatigue crack growth behavior of carbon black filled NR composites. *Composites Science and Technology*, **76**, 61-68.

44. Bhattacharya, M., Biswas, S., Bandyopadhyay, S., and Bhowmick, A. K. (2012) Influence of the nanofiller type and content on permeation characteristics of multifunctional NR nanocomposites and their modeling. *Polymers for Advanced Technologies*, **23**, 596-610.

45. Kim, W. S., Jang, S. H., Kang, Y. G., Han, M. H., Hyun, K., and Kim, W. (2013) Morphology and dynamic mechanical properties of styrene-butadiene rubber/silica/organoclay nanocomposites manufactured by a latex method. *Journal of Applied Polymer Science*, **128**, 2344-2349.

46. Shan, C., Gu, Z., Wang, L., Li, P., Song, G., Gao, Z., and Yang, X. (2011) Preparation, characterization, and application of NR/SBR/organoclay nanocomposites in the tire industry. *Journal of Applied Polymer Science*, **119**, 1185-1194.

47. Pal, K., Rajasekar, R., Pal, S. K., Kim, J. K., and Das, C. K. (2010) Influence of fillers on NR/SBR blends containing ENR-organoclay nanocomposites: Morphology and wear. *Journal of Nanoscience and Nanotechnology*, **10**, 3022-3033.

48. Tavakoli, M., Katbab, A. A., and Nazockdast, H. (2012) NR/SBR/organoclay nanocomposites: Effects of molecular interactions upon the clay microstructure and mechano-dynamic properties. *Journal of Applied Polymer Science*, **123**, 1853-1864.

49. Zarei, M., Naderi, G., Bakhshandeh, G., and Shokoohi, S. (2013) Ternary elastomer nanocomposites based on NR/BR/SBR: effect of nanoclay composition. *Journal of Applied Polymer Science*, **127**, 2038-2045.

50. Wang, Y., Liao, L., Zhong, J., He, D., Xu, K., Yang, C., Luo, Y., and Peng, Z. (2016) The behavior of natural rubber-epoxidized natural rubber-silica composites based on wet masterbatch technique. *Journal of Applied Polymer Science*, **133**, 10.1002/app.43571.

51. Utara, S., Jantachum, P., and Sukkaneewat, B. (2017) Effect of surface modification of silicon carbide nanoparticles on the properties of nanocomposites based on epoxidized natural rubber/natural rubber blends. *Journal of Applied Polymer Science*, **134**, doi:

10.1002/app.45289.

52. Manoharan, P., and Naskar, K. (2018) Plant based poly-functional hindered phenol derivative as an alternative for silane in cis-polyisoprene and its functionalized derivative systems containing nano-sized particulates. *Journal of Polymers and the Environment*, **26**(4), 1355-1370.

53. Fu, J.-F., Yu, W.-Q., Dong, X., Chen, L.-Y., Jia, H.-S., Shi, L.-Y., Zhong, Q.-D., and Deng, W. (2013) Mechanical and tribological properties of natural rubber reinforced with carbon blacks and Al_2O_3 nanoparticles. *Materials & Design*, **49**, 336-346.

54. Yang, G., Liao, Z., Yang, Z., Tang, Z., and Guo, B. (2015) Effects of substitution for carbon black with graphene oxide or graphene on the morphology and performance of natural rubber/carbon black composites. *Journal of Applied Polymer Science*, **132**, 10.1002/app.41832.

55. Sarkawi, S. S., Aziz, A. A., Aziz, A. K. C., Rahim, R. A., and Ismail, N. I. N. (2017) Properties of graphene nano-filler reinforced epoxidized natural rubber composites. *Journal of Polymer Science and Technology*, **2**, 36-44.

56. Lu, Y., Liu, J., Hou, G., Ma, J., Wang, W., Wei, F., and Zhang, L. (2016) From nano to giant? Designing carbon nanotubes for rubber reinforcement and their applications for high performance tires. *Composites Science and Technology*, **137**, 94-101.

57. Ismail, H., Ramly, A., and Othman, N. (2013) Effects of silica/multiwall carbon nanotube hybrid fillers on the properties of natural rubber nanocomposites. *Journal of Applied Polymer Science*, **128**, 2433-2438.

58. Poikelispaa, M., Das, A., Dierkes, W., and Vuorinen, J. (2013) The effect of partial replacement of carbon black by carbon nanotubes on the properties of natural rubber/butadiene rubber compound. *Journal of Applied Polymer Science*, **130**, 3153-3160.

59. Kumar, A., Modi, S., Panwar, A., Schmidt, D., Barry, C. M., and Mead, J. L. (2014) Highly impermeable nanocomposites of brominated butyl rubber with modified montmorillonite clay. *Rubber Chemistry and Technology*, **87**, 579-592.

60. Kumar, S., Nando, G., Nair, S., Unnikrishnan, G., Sreejesh, A., and Chattopadhyay, S. (2015) Effect of organically modified montmorillonite clay on morphological, physicomechanical, thermal stability, and water vapor transmission rate properties of BIIR-CO rubber nanocomposite. *Rubber Chemistry and Technology*, **88**, 176-196.

61. Halim, S., Lawandy, S., and Nour, M. (2013) Effect of in situ bonding system and surface modification of montmorillonite on the properties of butyl rubber/montmorillonite composites. *Polymer Composites*, **34**, 1559-1565.

62. He, F., Mensitieri, G., Lavorgna, M., de Luna, M. S., Filippone, G., Xia, H., Esposito, R., and Scherillo, G. (2017) Tailoring gas permeation and dielectric properties of bromobutyl rubber-graphene oxide nanocomposites by inducing an ordered nanofiller microstructure. *Composites, Part B: Engineering*, **116**, 361-368.

63. Xiong, X., Wang, J., Jia, H., Fang, E., and Ding, L. (2013) Structure, thermal conductivity, and thermal stability of bromobutyl rubber nanocomposites with ionic liquid modified graphene oxide. *Polymer Degradation and Stability*, **98**, 2208-2214.

64. Kotal, M., Banerjee, S. S., and Bhowmick, A. K. (2016) Functionalized graphene with polymer as unique strategy in tailoring the properties of bromobutyl rubber nanocomposites. *Polymer*, **82**, 121-132.

65. Frasca, D., Schulze, D., Bohning, M., Krafft, B., and Schartel, B. (2016) Multilayer graphene chlorine isobutyl isoprene rubber nanocomposites: influence of the multilayer graphene concentration on physical and flame-retardant properties. *Rubber Chemistry and Technology*, **89**, 316-334.

66. Frasca, D., Schulze, D., Wachtendorf, V., Krafft, B., Rybak, T., and Schartel, B. (2016) Multilayer graphene/carbon black/chlorine isobutyl isoprene rubber nanocomposites. *Polymers*, **8**, 95.

67. Kumar, V., Hanel, T., Giannini, L., Galimberti, M., and Giese, U. (2014) Graphene reinforced synthetic isoprene rubber nanocomposites. *KGK Rubberpoint*, *67*.

68. Zhang, Y., Mark, J. E., Zhu, Y., Ruoff, R. S., and Schaefer, D. W. (2014) Mechanical properties of polybutadiene reinforced with octadecylamine modified graphene oxide. *Polymer*, **55**, 5389-5395.

69. Lei, W., Zhou, X., Russell, T.P., Hua, K.-c., Yang, X., Qiao, H., Wang, W., Li, F., Wang, R., and Zhang, L. (2016) High performance bio-based elastomers: energy efficient and sustainable materials for tires. *Journal of Materials Chemistry A*, **4**, 13058-13062.

70. Qiao, H., Xu, W., Chao, M., Liu, J., Lei, W., Zhou, X., Wang, R., and Zhang, L. (2017) Preparation and performance of silica/epoxy group-functionalized biobased elastomer nanocomposite. *Industrial & Engineering Chemistry Research*, **56**, 881-889.

71. Sadasivuni, K. K., Saiter, A., Gautier, N., Thomas, S., and Grohens, Y. (2013) Effect of molecular interactions on the performance of poly (isobutylene-co-isoprene)/graphene and clay nanocomposites. *Colloid and Polymer Science*, **291**, 1729-1740.

72. Razzaghi-Kashani, M., Hasankhani, H., and Kokabi, M. (2007) Improvement in physical and mechanical properties of butyl rubber with montmorillonite organo-clay. *Iranian Polymer Journal*, **16**, 671-679.

73. Rajasekar, R., and Das, C. (2011) Development of butyl rubber nanocomposites in presence and absence of compatibiliser. *Plasti-*

cs, Rubber and Composites, **40**, 407-412.

74. Saritha, A., Joseph, K., Thomas, S., and Muraleekrishnan, R. (2012) Chlorobutyl rubber nanocomposites as effective gas and VOC barrier materials. *Composites, Part A: Applied Science and Manufacturing*, **43**, 864-870.

75. Azizli, M., Naderi, G., Bakhshandeh, G., Soltani, S., Askari, F., and Esmizadeh, E. (2014) Improvement in physical and mechanical properties of IIR/CR rubber blend organoclay nanocomposites. *Rubber Chemistry and Technology*, **87**, 10-20.

76. Azizli, M. J., Abbasizadeh, S., Hoseini, M., Rezaeinia, S., and Azizli, E. (2017) Influence of blend composition and organic cloisite 15A content in the structure of isobutylene-isoprene rubber/ethylene propylene diene monomer composites for investigation of morphology and mechanical properties. *Journal of Composite Materials*, **51**, 1861-1873.

77. Meera, A. P., Thomas, P., and Thomas, S. (2012) Effect of organoclay on the gas barrier properties of natural rubber nanocomposites. *Polymer Composites*, **33**, 524-531.

78. Yaragalla, S., Chandran, C. S., Kalarikkal, N., Subban, R., Chan, C. H., and Thomas, S. (2015) Effect of reinforcement on the barrier and dielectric properties of epoxidized natural rubber–graphene nanocomposites. *Polymer Engineering & Science*, **55**, 2439-2447.

79. Yari, H., Mohseni, M., Messori, M., and Ranjbar, Z. (2014) Tribological properties and scratch healing of a typical automotive nano clearcoat modified by a polyhedral oligomeric silsesquioxane compound. *European Polymer Journal*, **60**, 79-91.

80. Yari, H., Mohseni, M., and Messori, M. (2015) Toughened acrylic/melamine thermosetting clear coats using POSS molecules: Mechanical and morphological studies. *Polymer*, **63**, 19-29.

81. Yari, H., Mohseni, M., and Messori, M. (2016) A scratch resistant yet healable automotive clearcoat containing hyperbranched polymer and POSS nanostructures. *RSC Advances*, **6**, 76028-76041.

82. Yari, H., Moradian, S., and Tahmasebi, N. (2014) The weathering performance of acrylic melamine automotive clearcoats containing hydrophobic nanosilica. *Journal of Coatings Technology and Research*, **11**, 351-360.

83. Ramezanzadeh, B., Moradian, S., Khosravi, A., and Tahmasebi, N. (2011) A new approach to investigate scratch morphology and appearance of an automotive coating containing nano-SiO_2 and polysiloxane additives. *Progress in Organic Coatings*, **72**, 541-552.

84. Tahmassebi, N., Moradian, S., Ramezanzadeh, B., Khosravi, A., and Behdad, S. (2010) Effect of addition of hydrophobic nano silica on viscoelastic properties and scratch resistance of an acrylic/melamine automotive clearcoat. *Tribology International*, **43**, 685-693.

85. Mirabedini, S., Sabzi, M., Zohuriaan-Mehr, J., Atai, M., and Behzadnasab, M. (2011) Weathering performance of the polyurethane nanocomposite coatings containing silane treated TiO_2 nanoparticles. *Applied Surface Science*, **257**, 4196-4203.
86. Hang, T. T. X., Dung, N. T., Truc, T. A., Duong, N. T., Truoc, B. V., Vu, P. G., Hoang, T., Thanh, D. T. M., and Olivier, M.-G. (2015) Effect of silane modified nano ZnO on UV degradation of polyurethane coatings. *Progress in Organic Coatings*, **79**, 68-74.
87. Palimi, M., Rostami, M., Mahdavian, M., and Ramezanzadeh, B. (2014) Surface modification of Fe_2O_3 nanoparticles with 3-aminopropyltrimethoxysilane (APTMS): An attempt to investigate surface treatment on surface chemistry and mechanical properties of polyurethane/Fe_2O_3 nanocomposites. *Applied Surface Science*, **320**, 60-72.
88. Lahijania, Y. Z. K., Mohseni, M., and Bastani, S. (2014) Characterization of mechanical behavior of UV cured urethane acrylate nanocomposite films loaded with silane treated nanosilica by the aid of nanoindentation and nanoscratch experiments. *Tribology International*, **69**, 10-18.
89. Yahyaei, H., and Mohseni, M. (2013) Use of nanoindentation and nanoscratch experiments to reveal the mechanical behavior of sol–gel prepared nanocomposite films on polycarbonate. *Tribology International*, **57**, 147-155.
90. Verma, G., Kaushik, A., and Ghosh, A.K. (2013) Preparation, characterization and properties of organoclay reinforced polyurethane nanocomposite coatings. *Journal of Plastic Film & Sheeting*, **29**, 56-77.
91. Gurunathan, T., Mohanty, S., and Nayak, S.K. (2015) Effect of reactive organoclay on physicochemical properties of vegetable oil-based waterborne polyurethane nanocomposites. *RSC Advances*, **5**, 11524-11533.
92. Ghermezcheshme, H., Mohseni, M., and Yahyaei, H. (2015) Use of nanoindentaion and nanoscratch experiments to reveal the mechanical behavior of POSS containing polyurethane nanocomposite coatings: The role of functionality. *Tribology International*, **88**, 66-75.
93. Norouzi, S., Mohseni, M., and Yahyaei, H. (2016) Preparation and characterization of an acrylic acid modified polyhedral oligomeric silsesquioxane and investigating its effect in a UV curable coating. *Progress in Organic Coatings*, **99**, 1-10.
94. Nuraje, N., Khan, S.I., Misak, H., and Asmatulu, R. (2013) The addition of graphene to polymer coatings for improved weathering. *ISRN Polymer Science*, **2013**, Article ID 514617.
95. Liao, K.-H., Qian, Y., and Macosko, C. W. (2012) Ultralow percolation graphene/polyurethane acrylate nanocomposites. *Polymer*,

53, 3756-3761.

96. Miraftab, R., Ramezanzadeh, B., Bahlakeh, G., and Mahdavian, M. (2017) An advanced approach for fabricating a reduced graphene oxide-azo dye/polyurethane composite with enhanced ultraviolet (UV) shielding properties: Experimental and first-principles QM modeling. *Chemical Engineering Journal*, **321**, 159-174.

97. Dong, R., and Liu, L. (2016) Preparation and properties of acrylic resin coating modified by functional graphene oxide. *Applied Surface Science*, **368**, 378-387.

98. Al-Kawaz, A., Rubin, A., Badi, N., Blanck, C., Jacomine, L., Janowska, I., Pham-Huu, C., and Gauthier, C. (2016) Tribological and mechanical investigation of acrylic-based nanocomposite coatings reinforced with PMMA-grafted-MWCNT. *Materials Chemistry and Physics*, **175**, 206-214.

99. Kugler, S., Kowalczyk, K., and Spychaj, T. (2015) Hybrid carbon nanotubes/graphene modified acrylic coats. *Progress in Organic Coatings*, **85**, 1-7.

100. Kugler, S., Kowalczyk, K., and Spychaj, T. (2016) Influence of dielectric nanoparticles addition on electroconductivity and other properties of carbon nanotubes-based acrylic coatings. *Progress in Organic Coatings*, **92**, 66-72.

101. Motlagh, A. L., Bastani, S., and Hashemi, M. (2014) Investigation of synergistic effect of nano sized Ag/TiO_2 particles on antibacterial, physical and mechanical properties of UV-curable clear coatings by experimental design. *Progress in Organic Coatings*, **77**, 502-511.

102. Chouwatat, P., Nojima, S., Higaki, Y., Kojio, K., Hirai, T., Kotaki, M., and Takahara, A. (2016) An effect of surface segregation of polyhedral oligomeric silsesquioxanes on surface physical properties of acrylic hard coating materials. *Polymer*, **84**, 81-88.

103. Lee, Y.-H., Kim, H.-J., and Park, J.-H. (2013) Synthesis and characterization of polyester-based nanocomposites coatings for automotive pre-coated metal. *Progress in Organic Coatings*, **76**, 1329-1336.

104. Konwar, U., Mandal, M., and Karak, N. (2011) Mesua ferrea L. seed oil based acrylate-modified thermostable and biodegradable highly branched polyester resin/clay nanocomposites. *Progress in Organic Coatings*, **72**, 676-685.

105. Wang, Y.-J., Qiao, J., Baker, R., and Zhang, J. (2013) Alkaline polymer electrolyte membranes for fuel cell applications. *Chemical Society Reviews*, **42**, 5768-5787.

106. Kannan, R., Kakade, B. A., and Pillai, V. K. (2008) Polymer electrolyte fuel cells using Nafion-based composite membranes with functionalized carbon nanotubes. *Angewandte Chemie International Edition*, **47**, 2653-2656.

107. Farrukh, A., Ashraf, F., Kaltbeitzel, A., Ling, X., Wagner, M., Duran,

H., Ghaffar, A., Ur Rehman, H., Parekh, S. H., Domke, K. F., and Yameen, B. (2015) Polymer brush functionalized SiO$_2$ nanoparticle based Nafion nanocomposites: A novel avenue to low-humidity proton conducting membranes. *Polymer Chemistry*, **6**, 5782-5789.

108. Jia, W., Tang, B., and Wu, P. (2017) Novel composite proton exchange membrane with connected long-range ionic nanochannels constructed via exfoliated Nafion–boron nitride nanocomposite. *ACS Applied Materials & Interfaces*, **9**, 14791-14800.

109. Zhang, B., Cao, Y., Jiang, S., Li, Z., He, G., and Wu, H. (2016) Enhanced proton conductivity of Nafion nanohybrid membrane incorporated with phosphonic acid functionalized graphene oxide at elevated temperature and low humidity. *Journal of Membrane Science*, **518**, 243-253.

110. Peng, K.-J., Lai, J.-Y., and Liu, Y.-L. (2016) Nanohybrids of graphene oxide chemically-bonded with Nafion: Preparation and application for proton exchange membrane fuel cells. *Journal of Membrane Science*, **514**, 86-94.

111. Kowsari, E., Zare, A., and Ansari, V. (2015) Phosphoric acid-doped ionic liquid-functionalized graphene oxide/sulfonated polyimide composites as proton exchange membrane. *International Journal of Hydrogen Energy*, **40**, 13964-13978.

112. He, Y., Tong, C., Geng, L., Liu, L., and Lu, C. (2014) Enhanced performance of the sulfonated polyimide proton exchange membranes by graphene oxide: Size effect of graphene oxide. *Journal of Membrane Science*, **458**, 36-46.

113. Zhao, Y., Fu, Y., Hu, B., and Lu, C. (2016) Quaternized graphene oxide modified ionic cross-linked sulfonated polymer electrolyte composite proton exchange membranes with enhanced properties. *Solid State Ionics*, **294**, 43-53.

114. Gahlot, S., Sharma, P. P., Kulshrestha, V., and Jha, P. K. (2014) SGO/SPES-based highly conducting polymer electrolyte membranes for fuel cell application. *ACS Applied Materials & Interfaces*, **6**, 5595-5601.

115. Miao, S., Zhang, H., Li, X., and Wu, Y. (2016) A morphology and property study of composite membranes based on sulfonated polyarylene ether sulfone and adequately sulfonated graphene oxide. *International Journal of Hydrogen Energy*, **41**, 331-341.

116. Feng, M., Huang, Y., Cheng, T., and Liu, X. (2017) Synergistic effect of graphene oxide and carbon nanotubes on sulfonated poly(arylene ether nitrile)-based proton conducting membranes. *International Journal of Hydrogen Energy*, **42**, 8224-8232.

117. Oh, K.-H., Lee, D., Choo, M.-J., Park, K. H., Jeon, S., Hong, S. H., Park, J.-K., and Choi, J. W. (2014) Enhanced durability of polymer electrolyte membrane fuel cells by functionalized 2D boron nitride nanoflakes. *ACS Applied Materials & Interfaces*, **6**, 7751-7758.

118. Beydaghi, H., Javanbakht, M., and Kowsari, E. (2014) Synthesis and characterization of poly(vinyl alcohol)/sulfonated graphene oxide nanocomposite membranes for use in proton exchange membrane fuel cells (PEMFCs). *Industrial & Engineering Chemistry Research*, **53**, 16621-16632.

119. Liu, Y., Wang, J., Zhang, H., Ma, C., Liu, J., Cao, S., and Zhang, X. (2014) Enhancement of proton conductivity of chitosan membrane enabled by sulfonated graphene oxide under both hydrated and anhydrous conditions. *Journal of Power Sources*, **269**, 898-911.

120. Shirdast, A., Sharif, A., and Abdollahi, M. (2016) Effect of the incorporation of sulfonated chitosan/sulfonated graphene oxide on the proton conductivity of chitosan membranes. *Journal of Power Sources*, **306**, 541-551.

121. Bai, H., Li, Y., Zhang, H., Chen, H., Wu, W., Wang, J., and Liu, J. (2015) Anhydrous proton exchange membranes comprising of chitosan and phosphorylated graphene oxide for elevated temperature fuel cells. *Journal of Membrane Science*, **495**, 48-60.

122. Hidayah, N. M. S., Liu, W.-W., Lai, C. W., Noriman, N. Z., Sam, S. T., Hashim, U., and Lee, H.-C. (2016) Comparison of mechanical properties of graphene oxide and reduced graphene oxide based polymer composites. *Journal of Polymer Materials*, **33**, 615-628.

123. Shin, K.-Y., Hong, J.-Y., Lee, S., and Jang, J. (2012) Evaluation of anti-scratch properties of graphene oxide/polypropylene nanocomposites. *Journal of Materials Chemistry*, **22**, 7871-7879.

124. Milani, M. A., Gonzalez, D., Quijada, R., Basso, N. R., Cerrada, M. L., Azambuja, D. S., and Galland, G. B. (2013) Polypropylene/graphene nanosheet nanocomposites by in situ polymerization: synthesis, characterization and fundamental properties. *Composites Science and Technology*, **84**, 1-7.

125. Zhao, S., Chen, F., Zhao, C., Huang, Y., Dong, J.-Y., and Han, C. C. (2013) Interpenetrating network formation in isotactic polypropylene/graphene composites. *Polymer*, **54**, 3680-3690.

126. Bafana, A. P., Yan, X., Wei, X., Patel, M., Guo, Z., Wei, S., and Wujcik, E. K. (2017) Polypropylene nanocomposites reinforced with low weight percent graphene nanoplatelets. *Composites, Part B: Engineering*, **109**, 101-107.

127. Liang, J.-Z., Du, Q., Tsui, G. C.-P., and Tang, C.-Y. (2016) Tensile properties of graphene nano-platelets reinforced polypropylene composites. *Composites, Part B: Engineering*, **95**, 166-171.

128. Liang, J., Du, Q., Wei, L., Tsui, C., Tang, C., Law, W., and Zhang, S. (2015) Melt extrudate swell behavior of graphene nano-platelets filled-polypropylene composites. *Polymer Testing*, **45**, 179-184.

129. Ahmad, S. R., Xue, C., and Young, R. J. (2017) The mechanisms of reinforcement of polypropylene by graphene nanoplatelets. *Materials Science and Engineering B*, **216**, 2-9.

130. Ryu, S. H., and Shanmugharaj, A. (2014) Influence of long-chain alkylamine-modified graphene oxide on the crystallization, mechanical and electrical properties of isotactic polypropylene nanocomposites. *Chemical Engineering Journal*, **244**, 552-560.

131. Ryu, S. H., and Shanmugharaj, A. (2014) Influence of hexamethylene diamine functionalized graphene oxide on the melt crystallization and properties of polypropylene nanocomposites. *Materials Chemistry and Physics*, **146**, 478-486.

132. Yuan, B., Bao, C., Song, L., Hong, N., Liew, K. M., and Hu, Y. (2014) Preparation of functionalized graphene oxide/polypropylene nanocomposite with significantly improved thermal stability and studies on the crystallization behavior and mechanical properties. *Chemical Engineering Journal*, **237**, 411-420.

133. Quiles-Diaz, S., Enrique-Jimenez, P., Papageorgiou, D., Ania, F., Flores, A., Kinloch, I., Gómez-Fatou, M., Young, R., and Salavagione, H. (2017) Influence of the chemical functionalization of graphene on the properties of polypropylene-based nanocomposites. *Composites, Part A: Applied Science and Manufacturing*, **100**, 31-39.

134. Parameswaranpillai, J., Joseph, G., Shinu, K., Sreejesh, P., Jose, S., Salim, N. V., and Hameed, N. (2015) The role of SEBS in tailoring the interface between the polymer matrix and exfoliated graphene nanoplatelets in hybrid composites. *Materials Chemistry and Physics*, **163**, 182-189.

135. Parameswaranpillai, J., Joseph, G., Shinu, K., Jose, S., Salim, N. V., and Hameed, N. (2015) Development of hybrid composites for automotive applications: effect of addition of SEBS on the morphology, mechanical, viscoelastic, crystallization and thermal degradation properties of PP/PS-xGnP composites. *RSC Advances*, **5**, 25634-25641.

136. Idumah, C. I., Hassan, A., and Bourbigot, S. (2017) Influence of exfoliated graphene nanoplatelets on flame retardancy of kenaf flour polypropylene hybrid nanocomposites. *Journal of Analytical and Applied Pyrolysis*, **123**, 65-72.

137. Kim, M.-S., Yan, J., Kang, K.-M., Joo, K.-H., Kang, Y.-J., and Ahn, S.-H. (2013) Soundproofing ability and mechanical properties of polypropylene/exfoliated graphite nanoplatelet/carbon nanotube (PP/xGnP/CNT) composite. *International Journal of Precision Engineering and Manufacturing*, **14**, 1087-1092.

138. Ghasemi, F. A., Ghasemi, I., Menbari, S., Ayaz, M., and Ashori, A. (2016) Optimization of mechanical properties of polypropylene/talc/graphene composites using response surface methodology. *Polymer Testing*, **53**, 283-292.

139. Bouzouita, A., Notta-Cuvier, D., Raquez, J.-M., Lauro, F., and Dubois, P. (2017) Poly (lactic acid)-based materials for automotive applications. In: *Industrial Applications of Poly(Lactic acid)*, Di Lorenzo,

M. L., and Androsch, R. (eds.), Springer, Germany, pp. 177-219.

140. Notta-Cuvier, D., Murariu, M., Odent, J., Delille, R., Bouzouita, A., Raquez, J.M., Lauro, F., and Dubois, P. (2015) Tailoring polylactide properties for automotive applications: Effects of co-addition of halloysite nanotubes and selected plasticizer. *Macromolecular Materials and Engineering*, **300**, 684-698.

141. Notta-Cuvier, D., Odent, J., Delille, R., Murariu, M., Lauro, F., Raquez, J., Bennani, B., and Dubois, P. (2014) Tailoring polylactide (PLA) properties for automotive applications: Effect of addition of designed additives on main mechanical properties. *Polymer Testing*, **36**, 1-9.

142. Guo, J., Qiao, J., and Zhang, X. (2016) Effect of an alkalized-modified halloysite on PLA crystallization, morphology, mechanical, and thermal properties of PLA/halloysite nanocomposites. *Journal of Applied Polymer Science*, **133**, doi: 10.1002/app.44272.

143. Murariu, M., Dechief, A.-L., Peeterbroeck, S., Bonnaud, L., and Dubois, P. (2012) Polylactide (PLA)-halloysite nanocomposites: production, morphology and key-properties. *Journal of Polymers and the Environment*, **20**, 932-943.

144. Prashantha, K., Lecouvet, B., Sclavons, M., Lacrampe, M.F., and Krawczak, P. (2013) Poly(lactic acid)/halloysite nanotubes nanocomposites: structure, thermal, and mechanical properties as a function of halloysite treatment. *Journal of Applied Polymer Science*, **128**, 1895-1903.

145. Tham, W. L., Mohd Ishak, Z. A., and Chow, W. S. (2014) Mechanical and thermal properties enhancement of poly (lactic acid)/halloysite nanocomposites by maleic-anhydride functionalized rubber. *Journal of Macromolecular Science, Part B*, **53**, 371-382.

146. Tham, W. L., Poh, B. T., Mohd Ishak, Z. A., and Chow, W. S. (2014) Thermal behaviors and mechanical properties of halloysite nanotube-reinforced poly(lactic acid) nanocomposites. *Journal of Thermal Analysis and Calorimetry*, **118**, 1639-1647.

4

Living Polymer Architectures

4.1 Introduction

Living polymerization, also known as controlled polymerization, has become an important approach for the preparation of synthetic polymers having complex structures and controlled molecular weight distribution. Out of several living polymerization techniques, stable free radical polymerization, nitroxide mediated polymerization (NMP), reversible addition fragmentation chain transfer (RAFT), atom transfer radical polymerization (ATRP), etc., have proven to be the most noticeable methods for attaining controlled polymer architectures [1,2].

4.2 Living/Controlled Polymerization Techniques

In controlled polymerization, there is a possibility to reach a point of control on the resulting polymer, thus, achieving a required molecular weight with narrow distribution, if we are able to control the transfer of chain and chain termination during polymerization [3-5]. In such a scenario, the number of initiator groups equate to the number of polymer chains, thus, the process continues as long as the monomer is entirely consumed. The ends of the polymer chains stay active, and, hence, the polymerization can be continued further by adding more monomers [6-9].

4.2.1 Nitroxide Mediated Polymerization

This polymerization technique needs the addition of appropriate alkoxyamine to the polymerization process and has drawn a significant research interest owing to its simplicity. This method is based on the dissociation of dormant species and initiating alkoxyamine [10,11] reversibly, which avoids irreversible termination. Therefore, most of the dormant living chains may grow as long as the monomer molecules are available, thus, resulting in narrow molar mass distri-

Swati Singh and Vikas Mittal, Khalifa University of Science and Technology, Abu Dhabi, UAE

bution. It was first reported by Rizzardo [12] and Georges [13] that 2,2,6,6-tetramethyl-1-piperidynyloxyl nitroxide (TEMPO) can be used to control the polymerization of styrene. A large variety of nitroxides has been developed in order to expand the use for various monomers [14-22]. Boonpangrak *et al.* [23] also reported the synthesis of molecularly imprinted polymers (MIPs) via NMP. In the study, cholesterol-imprinted divinylbenzene (DVB)-based bulk MIPs were synthesized at 125 °C using cholesteryl(4-vinyl)phenyl carbonate as a template/functional monomer complex. Upon hydrolysis and removal of the template, the radio ligand binding experiments showed that the imprinted matrix obtained by NMP had an improved imprinting factor and binding affinity with respect to equivalent MIPs synthesized by free radical polymerization using benzoyl peroxide. Tomoeda *et al.* [24] polymerized n-butyl acrylate (BA) using emulsion nitroxide mediated polymerization. In this process at 100 °C, the authors used tetradecyltrimethylammonium bromide based cationic emulsifier and N-tert-butyl-N-(1-diethyl phosphono-2,2-dimethylpropyl) nitroxide. The molecular weight distribution of the as-developed polymers has been depicted in Figure 4.1.

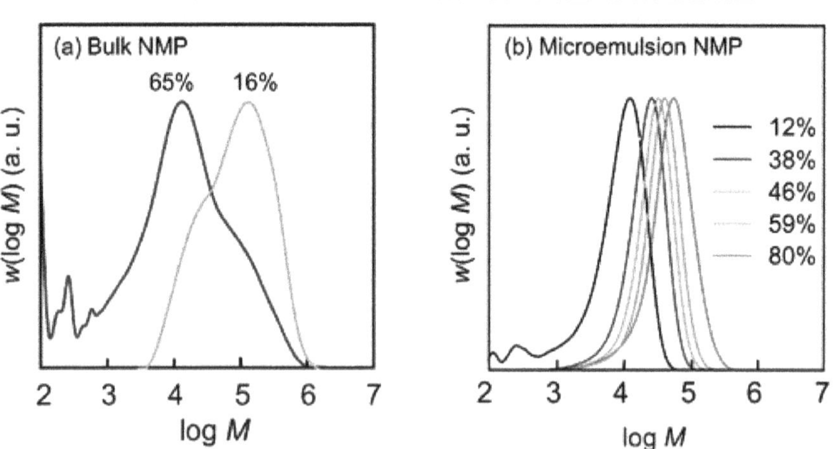

Figure 4.1 Molecular weight distributions (MWDs) at various conversions for (a) bulk (conversions: 16, 65%) and (b) microemulsion (conversions: 12, 38, 46, 59, 80%) NMPs of BA using SG1 at 100 °C: [AIBN]$_0$= 18.6mmol/L-monomer; [SG1]$_0$/[AIBN]$_0$ = 1.68, molar ratio. Reproduced from Reference 24 with permission from American Chemical Society.

4.2.2 Living Anionic and Telluride Mediated Polymerizations

Branched and linear polydiene structures can be synthesized using living anionic polymerization [25-30]. On the other hand, the living radical polymerization using organotellurium (TERP) can be used to synthesize polyacrylates [30-36]. Owing to the sensitivity of organomonotellurium compounds towards oxygen, it is important to provide inert atmosphere during synthesis. Azoinitiator and organoditelluride (DT) have been used by Yamago *et al.* [38], and DT was reported to be more stable than organomonotellurium compounds in ambient conditions [37,38]. Benedikt *et al.* [39] demonstrated telluride mediated polymerization method using benzoyl phenyltelluride as the reagent (Figure 4.2). It was observed that the material provided a low polydispersity in the range of 1.2-1.3. Mishima and Yamago [40] prepared copolymers comprising of methacrylates and α-alkenes using TERP technique. It was observed that the copolymers had a low polydispersity of less than 1.4. Other studies have reported similar use of TERP for generating living architectures [41-44].

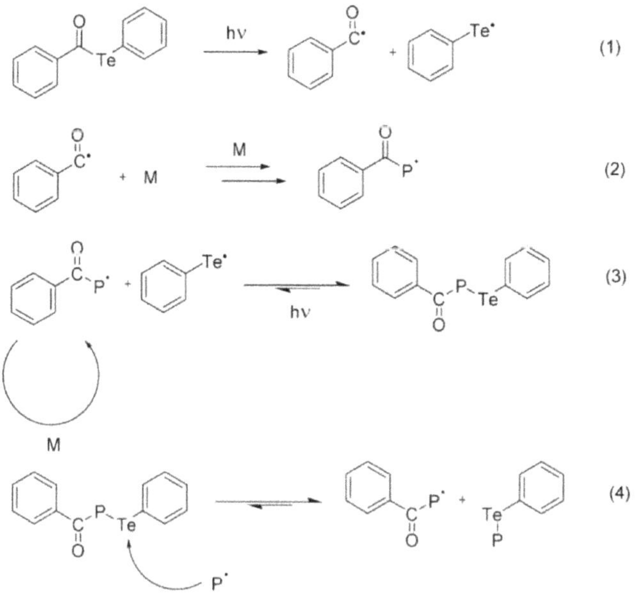

Figure 4.2 Schematic of living radical polymerization with benzoyl phenyltelluride as TERP-reagent. Reproduced from Reference 39 with permission from American Chemical Society.

4.2.3 Reversible Addition Fragmentation Chain Transfer

RAFT process combines the use of a chain transfer agent for achieving living structures [45-47]. Dithioesters [45] dithiocarbamates [48,49], trithiocarbonates [50] and xanthates [48,51] are the most widely used RAFT agents. Polymers with enhanced functionality and lower polydispersity index can be synthesized by choosing the suitable RAFT agent. For biomedical applications, RAFT polymers synthesized using biologically friendly solvents have also been explored. For instance, Ladavière *et al.* [52] performed the controlled polymerization of acrylic acid (AA) in alcohol and water, especially with phenoxyxanthates or with trithiocarbonate. Another study focused on the RAFT synthesis of poly(acrylic acid) in water with 4,4'-azobis(4-cyanovaleric acid) at 65 °C in the presence of water-soluble trithiocarbonate RAFT agent [53]. The resulting polymers were observed to have low dispersity ($\leq$1.2).

Ionizing radiation can be a method of choice for additive-free (i.e. initiator, accelerator, catalyst, etc.) synthesis of polymers at room temperature [54]. Hong *et al.* [55] used 60Co γ-rays for the initiation of the RAFT polymerization of AA and synthesized PAA with controlled molecular weight (M_n=4100-8800 g/mol) and dispersities (1.07-1.22). Millard *et al.* [56] obtained PAA with excellent control and without irreversible termination even at high conversion and high molecular weights (up to 143,000 g/mol) in the presence of S,S-bis(α,α'-dimethyl-α''-acetic acid)trithiocarbonate (TRITT) and 3-benzylsulfanylthiocarbonylsulfanyl propionic acid (BPATT) RAFT agents. Subsequently, the living character of PAA was confirmed by using it as a macrotransfer agent to synthesize block copolymers with acrylamide (AA_m), N-isopropylacrylamide (NIPAAm) and N,N-dimethylacrylamide (DMAAm) under γ-irradiation in aqueous media [57]. Klimkevicius and Makuska [58] reported successive RAFT polymerization of poly(ethylene oxide) methyl ether methacrylates with different length of PEO chains. Kulai *et al.* [59] described a novel array RAFT agents based on tin and evaluated the polymerization of styrene, methyl acrylate and acrylamides. ^{119}Sn NMR was reported to be a helpful tool to control the RAFT polymerization mediated by Sn (Figure 4.3). Moreover, it was observed that Sn-RAFT agents were highly reactive as anticipated from the electron donating triphenylstannyl analogs. This drove the RAFT polymerization of more activated monomers, but completely prevented the polymerization of less activated monomers.

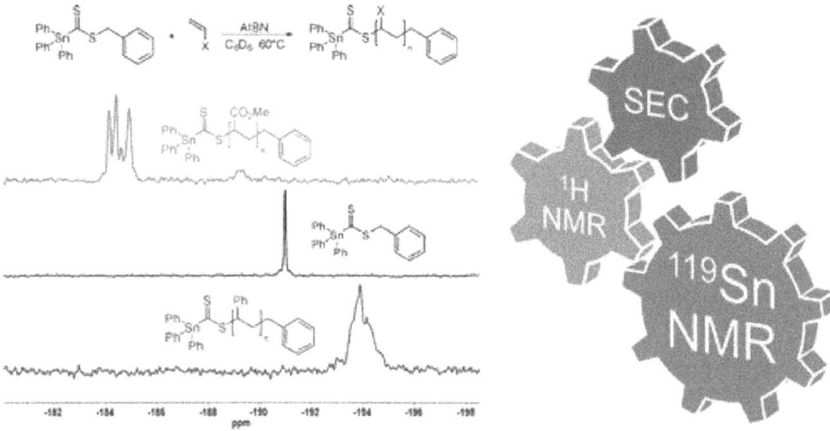

Figure 4.3 Schematic showing the use of ^{119}Sn NMR in RAFT polymerization. Reproduced from Reference 59 with permission from American Chemical Society.

Chaduc *et al.* [60] also studied the RAFT polymerization of methacyrlic acid in water. The polymer chains displayed a low dispersity (<1.19) in various conditions. Zhou *et al.* [61] studied seeded dispersion RAFT polymerization, poly(ethylene glycol) macro-RAFT agent mediated dispersion RAFT polymerization and seeded emulsion RAFT polymerization. It was observed that the dispersion RAFT polymerization proceeded slowly as compared to the other two. Also, the dispersion RAFT polymerization afforded vesicles, seeded dispersion RAFT polymerization led to the mixture of vesicles and porous nanospheres, whereas the seeded emulsion RAFT polymerization afforded porous nanospheres. Nicolay [62] also prepared copolymers of polythiol using RAFT polymerization method. In another study, Grover *et al.* [63] reported the preparation of polymers based on poly(PEGA) with pyridyl disulfide groups or aminooxy groups at the end of the chains by RAFT polymerization.

4.2.4 Atom Transfer Radical Polymerization

This polymerization technique is based on the transition metal facilitated atom transfer radical addition reactions [64-66]. The formation of propagating radicals is realized by the initiator, and a lower oxidation state transition metallic complex with halide coordinate [67] can terminate the process. Persistent radical effect can result in the reduction of termination as the reaction is progressing [68,69]. This results in increased chain length, viscosity and conversion [70]. As a consequence, there is a shift in the equilibrium towards the dormant species [69]. This process can be considered to be complex as it involves more than two oxidation states and a transition metal complex [71]. Equilibrium can be formed because of the different interactions between the species in the reaction medium [72,73]. The initiators usually used in the process are alkyl (pseudo)halides, which can be high or low molar mass compounds.

Shipp *et al.* [74] reported the preparation of copolymers containing methyl acrylate, methyl methacrylate (MMA) and butyl acrylate through atom transfer radical polymerization based on copper. A difunctional center block was developed following the addition of MMA to the two ends during the formation of PMMA-b-PBA-b-PMMA triblock copolymer. Figure 4.4 shows the molecular distribution of Br-PMA-Br macroinitiator and PMMA-b-PMA-b-PMMA

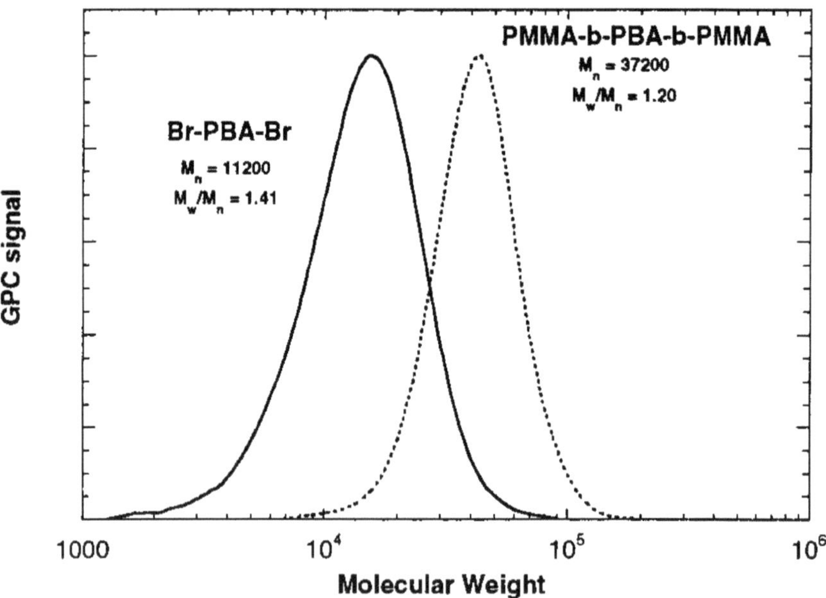

Figure 4.4 Molecular weight distributions of Br-PMA-Br macroinitiator and PMMA-b-PMA-b-PMMA triblock copolymer. Reproduced from Reference 74 with permission from American Chemical Society.

triblock copolymer. In another study, Robinson *et al.* [75] synthesized 2-hydroxyethyl methacrylate using ATRP at ambient temperature. ATRP and RAFT were combined for the synthesis of poly(vinyl acetate)-b-polystyrene (PVAc-b-PSt) with a well-defined structure by Tong *et al.* [76]. Successful implementation of activators regenerated by electron transfer atom transfer radical polymerization (ARGET ATRP) for aqueous media was achieved for the first time by Simakova *et al.* [77].

4.3 Advantages of Living Polymerization Techniques

Out of the various living polymerization techniques, RAFT offers some unique benefits, which include the capability to control the molecular weight and molecular weight distribution as well as the ability to resume the polymerization using dormant materials [78,79]. The various chain transfer agents used in RAFT polymerization provide increased acceptance towards different solvents and monomers [79,80-86]. Atom-transfer radical-polymerization is also known as a flexible living polymerization method, and it is possible to use this method for a wide range of monomers under suitable conditions [86,87]. However, this method is less appropriate for the monomers having acrylic groups and low reactivity [87,88]. The formation of graft and block copolymers is also effectively achieved in this method by forming halogenated polymer chains towards the end of the ATRP polymerization process [89,90]. The removal of metal catalysts is, however, tough after the ATRP process [91]. NMP has proved to be advantageous in many aspects even though the method is not suitable for situations requiring high temperature or long reaction time [92]. It is also not ideal for vinyl acetate based monomers [93]. The most significant benefit of NMP is the capability to generate highly pure polymers [93]. Also, NMP has been proven to be highly useful in the synthesis of polymers using 1,3-dienes [93]. NMP is also particularly valuable for the preparation of styrene [94,95].

4.4 Applications of Living Polymer Architectures

Living radical polymerization has driven a significant research interest in recent years owing to the potential to achieve "smart" multifunctional drug delivery methods. The function of proteins can be modified using beneficial gene delivery by replacing the genes which are not functional. An array of therapeutics based on nucleic acids have been observed, which can lead to possible treatments for infectious, hereditary and acquired syndromes. Plasmids are present in the candidate drugs which encourage small interfering RNAs and gene expression, which subsequently calms the target genes. Introduction of living free radical polymerization has brought an improvement in the control, quality and reproducibility of such materials. It is, thus, possible to achieve materials with required molecular weights and structures, along with narrow dispersion. Consequently, the impact of structure and molecular weight of the polymers on cytotoxicity has been analyzed, which is helpful to develop the scheme of next-generation vectors [96].

A control over density, chemical configuration, architecture and size make the polymer brushes an interesting topic of research. A layer of polymer attached to one side of the substrate surface and the other end hanging into solvent make the polymer brushes. The polymer chains stretch away from the surface because of both osmotic and steric repulsion between the chain segments in an appropriate solvent. Generally utilized artificial approach for developing polymer brushes is surface-initiated polymerization. Surface initiated ATRP offers the synthesis of precise polymer brushes owing to the benefits including simple preparation of initiator, good regulation of the chain growth, polymerization in aqueous solution and "living" end for copolymer embedding [97,98]. Other polymerization approaches explored to produce polymer brushes include RAFT [99], ring-opening metathesis polymerization [100] and NMP [101]. Also, the very precise control of the composition and construction due to living polymerization allows the improvement of various advanced materials with specific properties suitable for targeted applications, such as separations (chromatography and membrane), corrosion resistance, drug delivery, cell growth control, etc. Polymer brushes are also of particular interest in medical applications such as biomaterial implants and medical devices as well as diagnostics, packaging and blood contact materials [102,103]. Functional polymer brushes, especially the patterned and gradient architectures, are also employed extensively for biological applications, such as biosensing and biomolecule immobilization [104,105]. Polymer brush-assisted electroless deposition of metals (e.g., Cu, Ni, Ag or Au) on various substrates combined with lithography or printing techniques allows one to prepare inorganic/organic hybrid materials with potential applications in flexible electronics [106-108]. Applications in solar cells, memory storage and sensors have also been indicated [97].

4.5 Conclusion

In this chapter, different living polymerization techniques have been briefly reviewed. The multitude of controlled/living polymerization methods has helped to develop polymers with tuned structural characteristics. Owing to this, the developed polymers exhibit "living" nature and are useful for a wide variety of innovative applications.

References

1. Hasirci, V., Huri, P. Y, Tanir, T. E., Eke, G., Hasirci, N. (2017) Polymer fundamentals: polymer synthesis. *Comprehensive Biomaterials*, **1**, 349-371.
2. Piotti, M. E. (2001) Living cationic polymerization. In: *Encyclopedia of Materials: Science and Technology*, Elsevier, USA.
3. Calhoun, A., and Peacock, A. (2006) Polymer *Chemistry: Properties and Applications*, Hanser, Germany.
4. *High Modulus Polymers*, Zachariades A. E., and Porter R. S. (eds.), Marcel Dekker Inc., USA (1988).
5. Odian, G. (1991) *Principles of Polymerization*, 3ʳᵈ edition, Wiley, USA.
6. Szwarc, M. (1998) Living polymers. Their discovery, characterization, and properties. *Journal of Polymer Science, Part A: Polymer Chemistry*, **36**, 9-15.
7. Szwarc, M. (1956) 'Living' polymers. *Nature*, **178**, 1168-1169.
8. Xu, J., Jung, K., Atme, A., Shanmugam, S., and Boyer, C. (2014) A Robust and versatile photo induced living polymerization of conjugated and unconjugated monomers and its oxygen tolerance. *Journal of American Chemical Society*, **136**, 5508-5519.
9. Flory, P. (1953) *Principles of Polymer Chemistry*, Cornell University Press, USA.
10. He, J., Knoll, K., and McKee, G. (2004) Synthesis and properties of block copolymers of isoprene and 1,3-cyclohexadiene. *Macromolecules*, **37**, 4399-4405.
11. Goto, A., and Fukuda, T. (2004) Kinetics of living radical polymerization. *Progress in Polymer Science*, **29**, 329-385.
12. Solomon, D. H., Rizzardo, E., and Cacioli, P. (1986) Polymerization Process and Polymers Produced Thereby, patent US4581429.
13. Georges, M. K., Veregin, R. P. N., Kazmaier, P. M., and Hamer, G. K. (1993) Narrow molecular weight resins by a free-radical polymerization process. *Macromolecules*, **26**, 2987-2988.
14. Hong, K., and Mays, J. W. (2001) 1,3-Cyclohexadiene polymers. 1.

Anionic polymerization. *Macromolecules*, **34**, 782-786.

15. Grimaldi, S., Finet, J. P., Zeghdaoui, A., Tordo, P., Benoit, D., Gnanou, Y., Fontanille, M., Nicol, P., Pierson, J. F. (1997) *Polymer Preprints*, **38**(1), 651-652.

16. Benoit, D., Chaplinski, V., Braslau, R., Hawker, C. J. (1999) Development of a universal alkoxyamine for "living" free radical polymerizations. *Journal of American Chemical Society*, **121**, 3904-3920.

17. Benoit, D., Grimaldi, S., Robin, S., Finet, J. P., Tordo, P., and Gnanou, Y. (2000) Kinetics and mechanism of controlled free-radical polymerization of styrene and n-butyl acrylate in the presence of an acyclic β-phosphonylated nitroxide. *Journal of American Chemical Society*, **122**, 5929-5939.

18. Grimaldi, S., Lemoigne, F., Finet, J. P., Tordo, P., Nicol, P., and Plechot, M. (1996) Polymerization in the Presence of a β-Substituted Nitroxide Radical, patent WO1996024620A1.

19. Matyjaszewski, K. (2003) Controlled/living radical polymerization: State of the art in 2002. *ACS Symposium Series*, **854**, 2-9.

20. Sciannamea, V., Jerome, R., and Detrembleur, C. (2008) In-situ nitroxide-mediated radical polymerization (NMP) processes: their understanding and optimization. *Chemical Reviews*, **108**, 1104-1126.

21. Ide, N., and Fukuda, T. (1997) Nitroxide-controlled free-radical copolymerization of vinyl and divinyl monomers. Evaluation of pendant-vinyl reactivity. *Macromolecules*, **30**, 4268-4271.

22. Ide, N., and Fukuda, T. (1999) Nitroxide-controlled free-radical copolymerization of vinyl and divinyl monomers. 2. Gelation. *Macromolecules*, **32**, 95-99.

23. Boonpangrak, S., Whitcombe, M. J., Prachayasittikul, V., Mosbach, K., and Ye, L. (2006) Preparation of molecularly imprinted polymers using nitroxide-mediated living radical polymerization. *Biosensors and Bioelectronics*, **22**, 349-354.

24. Tomoeda, S., Kitayama, Y., Wakamatsu, J., Minami, H., Zetterlund, P. B., and Okubo M. (2011) Nitroxide-mediated radical polymerization in microemulsion (microemulsion NMP) of n-Butyl acrylate. *Macromolecules*, **44**, 5599-5604.

25. Shingo, K., Cheng, L., Thomas, R. H., and Marc, A. H. (2009) Controlled polymerization of a cyclic diene prepared from the ring-closing metathesis of a naturally occurring monoterpene. *Journal of American Chemical Society*, **131**, 7960-7961.

26. Hirao, A., Goseki, R., and Ishizone, T. (2014) Advances in living anionic polymerization: from functional monomers, polymerization systems, to macromolecular architectures. *Macromolecules*, **47**, 1883-1905.

27. Suzuki, T., Tsuji, Y., Takegami, Y., and Harwood, H. J. (1979) Micro-

structure of Poly(2-phenylbutadiene) prepared by anionic initiators. *Macromolecules*, **12**, 234-239.

28. (a) Inoue, H. I., Helbig, M., Vogl, O. (1977) Preparation and characterization of head-to-head polymers. 5. head-to-head polystyrene. *Macromolecules*, **10**, 1331-1339; (b) Akira, H., Yashunori, S., and Katsuhiko, T. (1998) Anionic living polymerization of 2,3-Diphenyl-1,3-butadiene. *Macromolecules*, **31**, 9141-9145; (c) Zhang, Y., Li, J., Li, X. H., He, J. P. (2014) Regio-specific polyacetylenes synthesized from anionic polymerizations of template monomers. *Macromolecules*, **47**, 6260-6269.

29. (a) Yuki, K., Keita, K., Sotaro, I., and Takashi, I. (2013) Living anionic polymerization of benzofulvene: highly reactive fixed transoid 1,3-Diene. *ACS Macro Letters*, **2**, 164-167; (b) Natori, I. (1997) Synthesis of polymers with an alicyclic structure in the main chain. living anionic polymerization of 1,3-cyclohexadiene with the n-butyllithium/n,n,n',n'-tetramethyl- ethylenediamine system. *Macromolecules*, **30**, 3696-3697.

30. Liu, K., He, Q., Ren, L., Gong, L. J., Hu, J. L., Ou, E. C., and Xu, W. J. (2016) Synthesis and characterization of the well-defined poly pentadiene via living anionic polymerization of (E)-1,3-pentadiene. *Polymer*, **89**, 28-40.

31. Yamago, S. (2006) Development of organotellurium-mediated and organostibine-mediated living radical polymerization reactions. *Journal of Polymer Science, Part A, Polymer Chemistry*, **44**, 1-12.

32. Yamago, S. (2009) Precision polymer synthesis by degenerative transfer controlled/living radical polymerization using organo tellurium, organostibine, and organobismuthine chain-transfer agents. *Chemical Reviews*, **109**, 5051-5068.

33. *Polymer Science: A Comprehensive Reference*, Matyjaszewski K., and Moller, M. (eds.), Elsevier, The Netherlands (2012).

34. Nakamura, Y., and Yamago, S. (2013) Organotellurium-mediated living radical polymerization under photoirradiation by a low-intensity light-emitting diode. *Polymer*, **54**, 981-994.

35. Zetterlund, P. B., Kagawa, Y., Okubo, M. (2008) Controlled/living radical polymerization in dispersed systems. *Chemical Reviews*, **108**, 3747-3794.

36. Huster, D., Yao, X., and Hong, M. (2002) Membrane protein topology probed by [1]h spin diffusion from lipids using solid-state NMR spectroscopy. *Journal of American Chemical Society*, **124**, 874-883.

37. Yamago, S., Iida, K., and Yoshida, J. (2002) Tailored synthesis of structurally defined polymers by organo tellurium-mediated living radical polymerization (TERP): synthesis of poly(meth)acrylate derivatives and their Di- and triblock copolymers. *Journal of American Chemical Society*, **124**, 13666-13667.

38. Yamago, S., Iida, K., Nakajima, M., Yoshida, J. (2003) Practical prot-

ocols for organo tellurium-mediated living radical polymerization by in situ generated initiators from AIBN and ditellurides. *Macromolecules*, **36**, 3793-3796.

39. Benedikt, S., Moszner, N., and Liska, R. (2014) Benzoyl phenyltelluride as highly reactive visible-light TERP reagent for controlled radical polymerization. *Macromolecules*, **47**, 5526-5531.

40. Mishima, E., and Yamago, S. (2011) Controlled alternating copolymerization of (meth)acrylates and vinyl ethers by using organo-heteroatom-mediated living radical polymerization. *Macromolecular Rapid Communications*, **32**(12), 893-898.

41. Sugihara, Y., Kagawa, Y., Yamago, S., and Okubo, M. (2007) Organotellurium-mediated living radical polymerization in miniemulsion. *Macromolecules*, **40**, 9208-9211.

42. Okubo, M., Sugihara, Y., Kitayama, Y., Kagawa, Y., and Minami, H. (2009) Emulsifier-free, organotellurium-mediated living radical emulsion polymerization of butyl acrylate. *Macromolecules*, **42**, 1979-1984.

43. Inui, T., Yamanishi, K., Sato, E., and Matsumoto, A. (2013) Organotellurium-mediated living radical polymerization (TERP) of acrylates using ditelluride compounds and binary azo initiators for the synthesis of high-performance adhesive block copolymers for on-demand dismantlable adhesion. *Macromolecules*, **46**, 8111-8120.

44. Sugihara, Y., Yamago, S., and Zetterlund, P. B. (2015) An innovative approach to implementation of organotellurium- mediated radical polymerization (TERP) in emulsion polymerization. *Macromolecules*, **48**, 4312-4318.

45. Chiefari, J., Chong, Y. K., Ercole, F., Krstina, J., Jeffery, J., Le, T. P. T., Mayadunne, R. T. A., Meijs, G. F., Moad, C. L., Moad, G., Rizzardo, E., and Thang, S. H. (1998) Living free-radical polymerization by reversible addition-fragmentation chain transfer: The RAFT process. *Macromolecules*, **31**, 5559-5562.

46. Moad, G., Rizzardo, E., and Thang, S. H. (2009) Living radical polymerization by the RAFT process – A second ipdate. *Australian Journal of Chemistry*, **62**, 1402-1472.

47. McCormick, C. L., Lowe, A. B. (2004) Aqueous RAFT polymerization: recent developments in synthesis of functional water-soluble (Co) polymers with Controlled Structures. *Accounts of Chemical Research*, **37**, 312-325.

48. Mayadunne, R. T. A., Rizzardo, E., Chiefari, J., Chong, Y. K., Moad, G., Thang, S. H. (1999) Living radical polymerization with reversible addition–fragmentation chain transfer (RAFT polymerization) using dithiocarbamates as chain transfer agents. *Macromolecules*, **32**, 6977-6980.

49. Destarac, M., Charmot, D., Franck, X., and Zard, S. Z. (2000) Dithio-

carbamates as universal reversible addition-fragmentation chain transfer agents. *Macromolecular Rapid Communications*, **21**, 1035-1039.

50. Mayadunne, R. T. A., Rizzardo, E., Chiefari, J., Kristina, J., Moad, G., Postma, A., Thang, S. H. (2000) Living polymers by the use of tri-thiocarbonates as reversible addition–fragmentation chain transfer (RAFT) agents: ABA triblock copolymers by radical polymerization in two steps. *Macromolecules*, **33**, 243-245.

51. Francis, R., and Ajayaghosh, A. (2000) Minimization of homopolymer formation and control of dispersity in free radical induced graft polymerization using xanthate derived macro-photoinitiators. *Macromolecules*, **33**, 4699-4704.

52. Ladaviere, C., Dorr, N., and Claverie, J. P. (2001) Controlled radical polymerization of acrylic acid in protic media. *Macromolecules*, **34**, 5370-5372.

53. Ji, J., Jia, L., Yan, L., and Bangal, P. R. (2010) Efficient synthesis of poly (acrylic acid) in aqueous solution via a RAFT process. *Journal of Macromolecular Science, Part A: Pure and Applied Chemistry*, **47**, 445-451.

54. Barsbay, M., and Guven, O. (2009) A short review of radiation-induced raft-mediated graft copolymerization: a powerful combination for modifying the surface properties of polymers in a controlled manner. *Radiation Physics and Chemistry*, **78**, 1054-1059.

55. Hong, C. Y., You, Y. Z., Bai, R. K., Pan, C. Y., and Borjihan, G. (2001) Controlled polymerization of acrylic acid under 60Co irradiation in the presence of dibenzyl trithiocarbonate. *Journal of Polymer Science, Part A: Polymer Chemistry*, **39**, 3934-3939.

56. Millard, P. E., Barner, L., Stenzel, M. H., Davis, T. P., Kowollik, C. B., Muller, A. H. (2006) RAFT polymerization of N-isopropylacrylamide and acrylic acid under γ-irradiation in aqueous media. *Macromolecular Rapid Communications*, **27**, 821-828.

57. Millard, P. E., Barner, L., Reinhardt, J., Buchmeiser, M. R., Kowollik, C. B., and Muller, A. H. (2010) Synthesis of water-soluble homo-and block-copolymers by RAFT polymerization under γ-irradiation in aqueous media. *Polymer*, **51**, 4319-4328.

58. Klimkevicius, V., and Makuska, R. (2017) Successive RAFT polymerization of poly (ethylene oxide) methyl ether methacrylates with different length of PEO chains giving diblock brush copolymers. *European Polymer Journal*, **86**, 94-105.

59. Kulai, I., Brusylovets, O., Voitenko, Z., Harrisson, S., Mazieres, S., and Destarac, M. (2015) RAFT polymerization with triphenyl stannyl carbodithioates (Sn-RAFT). *ACS Macro Letters*, **4**, 809-813.

60. Chaduc, I., Lansalot, M., D'Agosto, F., and Charleux, B. (2012) RAFT polymerization of methacrylic acid in water. *Macromolecules*, **45**, 1241-1247.

61. Zhou, H., Liu, C., Qu, Y., Gao, C., Shi, K., and Zhang, W. (2016) How the polymerization procedures affect the morphology of the block copolymer nanoassemblies: Comparison between dispersion RAFT polymerization and seeded RAFT polymerization. *Macromolecules*, **49**, 8167-8176.
62. Nicolay, R. (2012) Synthesis of well-defined polythiol copolymers by RAFT polymerization. *Macromolecules*, **45**, 821-827.
63. Grover, G. N., Lee, J., Matsumoto, N. M., and Maynard, H. D. (2012) Aminooxy and pyridyl disulfide telechelic poly(poly(ethylene glycol) acrylate) by RAFT polymerization. *Macromolecules*, **45**, 4958-4965.
64. Minisci, F. (1975) Free-radical additions to olefins in the presence of redox systems. *Accounts of Chemical Research*, **8**, 165-171.
65. Wang, J. S., and Matyjaszewski, K. (1995) Free-radical additions to olefins in the presence of redox systems. *Journal of American Chemical Society*, **117**, 5614-5615.
66. Matyjaszewski, K. (1998) Inner sphere and outer sphere electron transfer reactions in atom transfer radical polymerization. *Macromolecular Symposia*, **134**, 105-118.
67. Singleton, D. A., Nowlan, D. T., Jahed, N., and Matyjaszewski, K. (2003) Isotope effects and the mechanism of atom transfer radical polymerization. *Macromolecules*, **36**, 8609-8616.
68. Fischer, H. (2001) The persistent radical effect: a principle for selective radical reactions and living radical polymerizations. *Chemical Reviews*, **101**, 3581-3610.
69. Tang, W., Tsarevsky, N. V., and Matyjaszewski, K. (2006) Determination of equilibrium constants for atom transfer radical polymerization. *Journal of the American Chemical Society*, **128**, 1598-1604.
70. Vana, P., Davis, T. P., and Kowollik, C. B. (2002) Easy access to chain-length-dependent termination rate coefficients using RAFT polymerization. *Macromolecular Rapid Communications*, **23**, 952-956.
71. Matyjaszewski, K., Coca, S., Gaynor, S. G., Wei, M., and Woodworth, B. E. (1998) Controlled radical polymerization in the presence of oxygen. *Macromolecules*, **31**, 5967-5969.
72. Braunecker, W. A., and Matyjaszewski, K. (2007) Controlled/living radical polymerization: Features, developments, and perspectives. *Progress in Polymer Science*, **32**, 93-146.
73. Tsarevsky, N. V., Braunecker, W. A., and Matyjaszewski, K. (2007) Electron transfer reactions relevant to atom transfer radical polymerization. *Journal of Organometallic Chemistry*, **692**, 3212-3222.
74. Shipp, D. A., Wang, J. L., and Matyjaszewski, K. (1998) Synthesis of acrylate and methacrylate block copolymers using atom transfer radical polymerization. *Macromolecules*, **31**, 8005-8008.

75. Robinson, K. L., Khan, M. A., Banez, M. V. P., Wang, X. S., and Armes, S. P. (2001) Controlled polymerization of 2-hydroxyethyl methacrylate by ATRP at ambient temperature. *Macromolecules*, **34**, 3155-3158.

76. Tong, Y. Y., Dong, Y. Q., Du, F. S., and Li, Z. C. (2008) Synthesis of well-defined poly(vinyl acetate)-b-polystyrene by combination of ATRP and RAFT polymerization. *Macromolecules*, **41**, 7339-7346.

77. Simakova, A., Averick, S. E., Konkolewicz, D., and Matyjaszewski, K. (2002) Aqueous ARGET ATRP. *Macromolecules*, **45**, 6371-6379.

78. Moad, G., and Solomon, D. H. (2006) *The Chemistry of Radical Polymerization*, 2nd edition, Elsevier, USA.

79. Moad, G., Chiefari, J., Chong, Y. K., Krstina, J., Mayadunne, R. T. A., Postma, A., Rizzardo, E., and Thang, S. H. (2000) Living free radical polymerization with reversible addition–fragmentation chain transfer (the life of RAFT). *Polymer International*, **49**, 993-1001.

80. Daikh, B. E., and Finke, R. G. (1992) The persistent radical effect: a prototype example of extreme, 105 to 1, product selectivity in a free-radical reaction involving persistent .cntdot.CoII[macrocycle] and alkyl free radicals. *Journal of the American Chemical Society*, **114**, 2938-2943.

81. Boyer, C., Bulmus, V., Davis, T. P., Ladmiral, V., Liu, J. Q., and Perrier, S. (2009) Bio applications of RAFT polymerization. *Chemical Reviews*, **109**, 5402-5436.

82. Monteiro, M. J. (2005) Modeling the molecular weight distribution of block copolymer formation in a reversible addition–fragmentation chain transfer mediated living radical polymerization. *Journal of Polymer Science, Part A: Polymer Chemistry*, **43**, 3189-3204.

83. Gossage, R. A., Kuil, L. A. V., and Koten, G. V. (1998) Diaminoarylnickel(II) "Pincer" complexes: mechanistic considerations in the kharasch addition reaction, controlled polymerization, and dendrimeric transition metal catalysts. *Accounts of Chemical Research*, **31**, 423-431.

84. Moad, G., Rizzardo, E., and Thang, S. H. (2006) Living radical polymerization by the RAFT process - A first update. *Australian Journal of Chemistry*, **59**, 669-692.

85. Moad, G., Ching, Y. K., Postma, A., Rizzardo, E., and Thang, S. H. (2005) Advances in RAFT polymerization: the synthesis of polymers with defined end-groups. *Polymer*, **46**, 8458-8468.

86. Derry, M. J., Fielding, L. A., and Armes, S. P. (2016) Polymerization-induced self-assembly of block copolymer nanoparticles via RAFT non-aqueous dispersion polymerization. *Progress in Polymer Science*, **52**, 1-18.

87. Ayres, N. (2011) Atom transfer radical polymerization: a robust and versatile route for polymer synthesis. *Polymer Reviews*, **51**,

138-162.

88. Ran, J., Wu, L., Zhang, Z., and Xu, T. (2014) Atom transfer radical polymerization (ATRP): A versatile and forceful tool for functional membranes. *Progress in Polymer Science*, **39**, 124-144.

89. Patten, T. E., and Matyjaszewski, K. (1998) Atom transfer radical polymerization and the synthesis of polymeric materials. *Advanced Materials*, **10**, 901-915.

90. Krys, P., and Matyjaszewski, K. (2017) Kinetics of atom transfer radical polymerization. *European Polymer Journal*, **89**, 482-523.

91. Ouchi, M., Terashima, T., and Sawamoto, M. (2008) Precision control of radical polymerization via transition metal catalysis: from dormant species to designed catalysts for precision functional polymers. *Accounts of Chemical Research*, **41** (9), 1120-1132.

92. Hawker, C. J., Bosman, A. W., and Harth, E. (2001) New polymer synthesis by nitroxide mediated living radical polymerizations. *Chemical Reviews*, **101**, 3661-3688.

93. Grubbs, R. B. (2011) Nitroxide-mediated radical polymerization: limitations and versatility. *Polymer Reviews*, **51**, 104-137.

94. Beejapur, H. A., Zhang, Q., Hu, K., Zhu, L., Wang, J., and Ye, Z. (2019) TEMPO in chemical transformations: From homogeneous to heterogeneous. *ACS Catalysis*, **9**(4), 2777-2830.

95. Chong, Y. K., Ercole, F., Moad, G., Rizzardo, E., Thang, S. H., and Anderson, A. G. (1999) Imidazolidinone nitroxide-mediated polymerization. *Macromolecules*, **32**, 6895-6903.

96. Chu, D. S. H., Schellinger, J. G., Shi, J., Convertine, A. J., Stayton, P. S., and Pun, S. H. (2012) Application of living free radical polymerization for nucleic acid delivery. *Macromolecules*, **45**, 1089-1099.

97. Li, B., Yu, B., Ye, Q., and Zhou, F. (2015) Tapping the potential of polymer brushes through synthesis. *Accounts of Chemical Research*, **48**, 229-237.

98. Zhou, F., Zheng, Z., Yu, B., Liu, W., Huck, W. T. S. (2006) Multicomponent polymer brushes. *Journal of American Chemical Society*, **128**, 16253-16258.

99. Baum, M., Brittain, W. J. (2002) Synthesis of polymer brushes on silicate substrates via reversible addition fragmentation chain transfer technique. *Macromolecules*, **35**, 610-615.

100. Ye, Q., Wang, X., Li, S., and Zhou, F. (2010) Surface-initiated ring-opening metathesis polymerization of pentadecafluorooctyl-5-norbornene-2- carboxylate from variable substrates modified with sticky biomimic initiator. *Macromolecules* **43**, 5554-5560.

101. Husseman, M., Malmstro, E. E., McNamara, M., Mate, M., Mecerreyes, D., Benoit, D. G., Hedrick, J. L., Mansky, P., Huang, E., Russell, T. P., and Hawker, C. J. (1999) Controlled synthesis of polymer brushes by "living" free radical polymerization techniques. *Macromolecules*, **32**, 1424-1431.

102. Wei, Q., Cai, M., Zhou, F., and Liu, W. (2013) Dramatically tuning friction using responsive polyelectrolyte brushes. *Macromolecules*, **46**, 9368-9379.

103. Wei, Q., Becherer, T., Uberti, S. A., Dzubiella, J., Wischke, C., Neffe, A. T., Lendlein, A., Ballauff, M., and Haag, R. (2014) Protein interactions with polymer coatings and biomaterials. *Angewandte Chemie International Edition*, **53**, 8004-8031.

104. Welch, M. E., Ritzert, N. L., Chen, H., Smith, N. L., Tague, M. E., Xu, Y., Baird, B. A., Abruna, H. D., and Ober, C. K. (2014) Generalized platform for antibody detection using the antibody catalyzed water oxidation pathway. *Journal of American Chemical Society*, **136**, 1879-1883.

105. Guo, R., Yu, Y., Xie, Z., Liu, X., Zhou, X., Gao, Y., Liu, Z., Zhou, F., Yang, Y., and Zheng, Z. (2013) Matrix-assisted catalytic printing for the fabrication of multiscale, flexible, foldable, and stretchable metal conductors. *Advanced Materials*, **25**, 3343-3350.

106. Jiang, H., and Xu, F. J. (2013) Biomolecule-functionalized polymer brushes. *Chemical Society Reviews*, **42**, 3394-3426.

107. Paripovic, D., and Klok, H. A. (2011) polymer brush guided formation of thin gold and palladium/gold bimetallic films. *ACS Applied Materials & Interfaces*, **3**, 910-917.

108. Wang, X., Guo, Q., Cai, X., Zhou, S., Kobe, B., and Yang, J. (2014) Initiator- integrated 3D printing enables the formation of complex metallic architectures. *ACS Applied Materials & Interfaces*, **6**, 2583-2587.

5

Epoxy: Understanding Degradation and Stabilization

5.1 Introduction

Epoxy resins are either low molecular weight pre-polymers or cured polymeric networks. Historically, epoxy compounds have been discovered by Prileschaiev in 1909, and the commercial products including adhesives and casting resins were marketed in USA in 1946 [1,2]. Epoxy resins exhibit outstanding features such as high thermal and chemical stability as well as adhesion and flame resistance depending on the processing method. Nowadays, epoxies are one of the most widely used thermosetting resins in the fields of adhesives, coatings, semiconductor encapsulation, hardware components, electronic circuit board materials and aerospace, along with other engineering fields.

Polymer degradation is a critical issue as it causes the discoloration, embrittlement and material deterioration [3]. Thermal degradation, oxidation and UV irradiation are the major routes for the degradation processes. The degradation of epoxy takes place largely under high temperature and oxidative conditions. Exposure of UV light combined with oxygen leads to a synergetic effect. Stabilization process depends on the type of stabilizers that act as hydroperoxide decomposers, radical scavengers, UV absorbers and reaction modifiers such as fire retardants. The stabilizer usually reacts in amorphous solid phase or highly viscous state [4-6]. The mobility of the stabilizer is required to be enough to reach the reaction site, but not too high to be lost through the surface by migration. The molecular weight of the commercial stabilizers is usually adjusted in the range of 200-2000 g/mol.

In this chapter, the degradation behavior of epoxy is briefly studied by focusing on the mechanism of induced damage. Stabilization of epoxy with additives to overcome the degradation phenomena is also briefly introduced together. Such insights into the degradation and stabilization phenomena are required for expanding the commercial uses of epoxy.

Hyemin Lee, Zhou He, Yuting Li and Vikas Mittal, Khalifa University of Science and Technology, Abu Dhabi, UAE

5.2 Understanding the Degradation and Stabilization of Epoxy

5.2.1 Thermal Degradation

Two major factors have an effect on the thermal degradation of epoxy: the structure of epoxy and service conditions. Monomer structure and curing agent used to generate the network determine the thermal behavior. Polymers with aromatic rings in the structure are generally considered to be thermally stable. Similarly, the aromatic epoxy resin is much more stable than the aliphatic one because of the ring structure. Curing agent is generally adopted by considering application conditions, workability and other factors. Christiansen *et al.* [7] revealed that the epoxy resins cured with novolacs had excellent thermal stability, while the epoxy resins cured with dicyandiamide (DICY) performed poorly. Regarding service conditions, epoxy has overall high thermal stability. Degradation generally initiates above 150 °C. The investigation of the thermal ageing stability of epoxy has been mostly performed at high temperatures accordingly. It has been reported that the unfilled epoxy shows short-term thermal resistance at 220-250 °C and rapid degradation over 250 °C. The duration of the exposure of heat and partial oxygen pressure also contributes to the thermal degradation. Also, it is important to notice that the thermal degradation of epoxy resins at relatively low temperature is controlled by diffusion rather than reaction.

Thermal Degradation Mechanism of Epoxy

Degradation can proceed by chain scission and crosslinking, depending on the characteristics of epoxy [8]. It has been understood that chain scission usually contributes to degradation, whereas crosslinking can enhance stability. One of the indicators to assess the degradation mechanism is the glass transition temperature (T_g). The extent of chain scission can be deduced by measuring the decrease of T_g. Reversely, the crosslinking development in the molecule is assumed when T_g increases upon thermal ageing.

Thermal degradation of the epoxy resins at high temperatures is generally initiated by the formation of vinylene ethers on the backbone when secondary alcohol is dehydrated [9]. The dehydration causes the allylic ether bond to be weakened, which leads to chain scission in the structure. Depending on the curing agent, the stability of the bond varies. The epoxy resins cured by anhydrides are observed to have better thermal stability than the epoxy resins cured by amines. It is assumed that the nucleophilic nature of the amine groups within the network leads to a weak allylic amine C-N bond as compared to the C-O bond. Amine tends to undergo either volatilization or charring as solid residue as the following step. Secondary chain scission of the amine group (C-NH) has been rarely observed in the amine-cured epoxy resins [10].

The study by Zahra *et al.* [11] introduced the possible chain scission route for diglycidyl ether of bisphenol A (DGEBA) cured by jeffamine (POPA) or polyamidoamine (PAA) in low temperature range from 70 to 150 °C by analyzing the IR spectra. It was suggested that the amide group was formed as a major product of oxidation induced by the attack of α amino carbon group of DGEBA, regardless of the curing agents. In the long term experiment for almost a month at 150 °C, the T_g of DGEBA cured by POPA kept decreasing, while the T_g of DGEBA cured by PAA showed increasing tendency, which could be interpreted as predominant chain scission with DGEBA/POPA system and predominant crosslinking reaction with DGEBA/PAA system with time.

Ernault *et al.* [12] tested the steric effect of diamine hardener on the thermal oxidation mechanism of the DGEBA based epoxy resin. The oxidation mechanism of DGEBA resins cured with cycloaliphatic (isophorone diamine; IPDA) or linear aliphatic (trioxa-tridecanediamine or TTDA) diamine hardener was deduced by tracking the T_g variations. It was observed that DGEBA/IPDA resin mainly took the chain scission route with faster oxidation rate as compared to EGEBA/TTDA system that presumably underwent crosslinking in some parts.

In the case of epoxy-clay nanocomposites, it is suggested that the thermal degradation behavior of the DGEBA-montmorillonite (MMT) composites relies on mainly three factors, including the clay loading, structure of nanocomposite and purge gas [13]. Pandey *et al.* [14], however, argued that MMT could influence in both positive and negative ways as the reactivity of the epoxy resin with the curing agent decreased when MMT was added, whereas MMT conferred insulating function to impart protection from oxygen [14]. For example, the exfoliated nanocomposites exhibit better barrier properties and thermal stability than the corresponding intercalated ones as the interference effect is dominant for the intercalated nanocomposites, while the insulation effect is observed to be largely dominant for the exfoliated ones [15].

Thermo-oxidative Degradation Behavior of Epoxy

Degradation behavior of epoxy under thermo-oxidative conditions has been widely studied. Intrinsic characteristics of epoxy networks including the monomer structure and curing agent play an important role in the thermo-oxidation of epoxy. It is also important to notice that the degradation is highly diffusion limited [16]. Discoloration and the formation of oxidized layer with weight loss and density decrease on the surface of epoxy appear as a result, thus, leading the surface layer to shrink and possibly crack [17].

Other reports have pointed out that the oxidative reaction in epoxy resins is limited to the surface as the oxidized layer provides protection from further oxidation. One recent study concluded that the embrittlement in the amine-cured epoxy resin was not directly related to T_g or weight loss, but was governed by the oxidation of hydroxypropylether groups in the DGEBA segment, regardless of the curing agent [18]. As per the study, the gap between T_g and imposed temperature did not show relevance with embrittlement, but the network undergoing crosslinking exhibited significant mass loss [18].

Several recent studies have reported the superficial oxidized layer development on epoxy materials [19-21]. For instance, Yang *et al.* [21] studied the oxidation of the anhydride-cured epoxy resin under thermal conditions (130-160 °C) for 30 days and observed noticeable free volume fraction decrease in the skin layer relative to the bulk sample. The oxidized skin thickness was observed to be less than 100 μm. The comparison of the compressive behavior of the epoxy resin and 3D carbon fibers/epoxy braided composite was also performed after thermal aging by Zhang *et al.* [20]. Stress-strain curves of the resin and 3D composite indicated that the curve of the epoxy resin became stable after 1 day, however, the composite showed continuous decrease with time. It could be assumed that the epoxy resin was protected by the oxidized layer on the surface, while the interface cracking in the composite led to crack propagation and oxygen penetration into the core.

Thermo-oxidative reaction of the epoxy resins under gamma irradiation has received attention because of the application in nuclear industry as coatings. Most of recent studies have focused on the oxidation reaction of epoxy with relatively low glass transition temperatures (typically 50 °C $\leq T_g \leq$ 80 °C). Galant *et al.* [22] studied the thermal and radio-oxidation of the epoxy coating with low T_g epoxy-amine networks under low temperature conditions (50-110 °C) at

200 Gy/h and observed that the tertiary amine group was vulnerable to oxidation regardless of the type of exposure (thermal or radio), thus, resulting in the amide group formation. Variation of the gamma ray dosage accelerated the oxidation rate at low level (50 Gy/h), but the rate became constant at high dosage (200 and 2000 Gy/h). Comparison of the DGEBA/PAA and DGEBA/POPA systems at low level of irradiation indicated that the DGEBA/POPA system developed cross-linking, while the DGEBA/PAA system went through chain scission reaction [23]. Another degradation study of the epoxy-amine network showed carbonyl and hydroperoxide formation under oxygen and gamma irradiation conditions [24]. The study also reported that chain scission was predominant in oxidative degradation.

5.2.2 Fire Retardation

Highly flammable nature of the epoxy based materials has been an issue for extended applications in many fields such as electronics, automotives, aircrafts, etc. The decomposition of the epoxy resins is affected by the reactive curing agents and fire retardant additives. Reactive curing agents such as amines, anhydrides and phenolic groups have a significant impact on flammability. Generally, the amine-cured epoxy resins show better flame stability as compared to the anhydride-cured epoxy resins. The ratio of carbon to oxygen in the structure is closely linked to the combustion behavior. Regarding the inorganic fillers, such as alumina or silica, Paterson-Jones *et al.* [25] also suggested that the catalytic reaction by the fillers in the epoxy resin could accelerate thermal decomposition.

The decomposition mechanism of brominated epoxy resin (DGEBA/DDS) was suggested by Luda *et al.* [26]. It was observed that the brominated part decomposed first, followed by the decomposition of the other parts and char formation. Halogenated compounds are, however, substituted by alternatives because of the toxic gas generation issue. The phosphorous compounds have received attention to replace the halogenated compounds as fire retardants. It is suggested that the flame retardant characteristic is provided by the phosphorous compounds through modification of the thermal decomposition during ignition, which yields carbonaceous char rather than CO and CO_2. The char acts as a physical barrier for combustible gases as well as thermal insulation in the decomposition process. The study by Liu *et al.* [27] proposed that the thermal degradation mechanism of phosphorous containing epoxy (bis-(3-glycidyloxy) phenylphosphine oxide   (BGPPO) and 4,4'-diaminodiphenylsulfone (DDS) initiated from phosphorous-phenol bond. Dehydration by P-O-C chain scission and elimination of propyl groups took place during degradation, subsequently leading to char yields with high phosphorous contents.

5.2.3 UV Degradation

Epoxy coatings are widely used as long-term protective layers to safeguard the metal surfaces due to high mechanical properties, adhesion and solvent resistance. The coatings are often exposed to oxidative and UV irradiation conditions, which can lead to the degradation of the material. Many research studies have been carried out to understand the degradation mechanism under UV, such as pyrolysis or photo-oxidation, and to establish the pathways to improve the light resistance of epoxy.

The photo-oxidation mechanism of epoxy maleate of bisphenol A (EMBA) through hydroperoxide intermediates has been suggested by Rosu *et al.* [28]. The degradation was observed to be initiated by the free radical reaction and excitation of double bonds in EMBA at initial stage. Some crosslinked structure formation in EMBA was also suspected at an early stage of photolysis. Hydrogen peroxide formation as secondary reaction led to chain scission in the structure, producing etheric aromatic group.

In contrast, the properties of epoxy thermosets can be improved in many cases at initial stages of UV exposure owing to the residual crosslinking before degradation [29]. Liu *et al.* [30] studied the microstructure evolution of the epoxy coatings, polyamide-cured DGEBA, under UV irritation (55 mW/cm²) for 399 h and revealed that the epoxy coatings could be reinforced under UV light. Upon exposure, the development of a highly crosslinked structure was observed on the surface of the epoxy coatings as a result of recombination, which enhanced the water resistance of the coatings.

5.2.4 Epoxy Stabilization

Epoxy has a wide range of applications. The stability and non-flammability of epoxy at the processing temperature and outdoor condition are essential properties. The purpose of stabilization is to conserve the original properties during various service conditions. Stabilizers are, thus, added to slow down the deterioration, discoloration and embrittlement of epoxy [3]. Stabilizers are mostly dispersed in the amorphous region, and the morphology affects the stabilization process. The stabilizer optimization is of immense importance as excess mobility can lead to the loss of the material. Frequently used stabilizer types for epoxy are antioxidants, light stabilizers, fire retardants, etc.

Antioxidants

Oxidation is one of the major processes leading to polymer degradation. Antioxidants function by combining with free radicals to prevent the progress of degradation. Low molecular weight compounds are usually employed as free radical scavengers. However, majority of the research studies on epoxy materials have focused on the thermal decomposition and development of flame retardants, rather than antioxidants. Recently, eugenol-based epoxy materials have been designed as effective antioxidant [31]. Burton [32] also studied the protective effect of the commercial antioxidants such as tris(nonylpheny1) phosphite on triethylenetetraamin-cured DGEBA at 125 °C. No noticeable protection was observed, thus, explaining that the predominant degradation mechanism of the aliphatic amine-cured epoxy resins was not controlled by free radical processes.

Ghasemi-Kahrizsangi *et al.* [33] investigated the corrosion behavior of the modified nano-carbon black (CB)/epoxy coatings at 65 °C. Addition of CB in the coatings resulted in better corrosion resistance owing to the improved barrier against ion diffusion. Wei *et al.* [34] also applied carbon black to obtain protection for mild steel in 3% NaCl. CB-filled fusion bonded epoxy coatings were observed to affect the corrosion behavior in a positive way (with CB weight fraction above percolation threshold).

Light Stabilizers

Light stabilizers include UV absorbers, quenchers of excited states, hydroperoxide decomposers and free radical scavengers. Many materials are observed to have several of such functions, e.g. carbon black. Light stabilizers interfere with the light-induced reactions and achieve inhibition. Dispersion, particle size and type of light stabilizers are important factors affecting their function [35].

Carbon black is known for its effective UV stabilizing efficiency as well as antioxidant effect in the polymer industry [36,37]. CB/epoxy nanocomposite coatings were investigated by Ghasemi-Kahrizsangi *et al.* [38] under UV irradiation to investigate the effect of CB nanoparticles. CB interfered with the formation of micro-cracks and penetration of ionic species by absorbing UV. The epoxy coatings with 2.5 wt% CB exhibited effective degradation resistance after 2000 h exposure and 24 h immersion in 3.5 wt% NaCl.

The addition of clay in epoxy is often observed to improve thermal, viscoelastic and mechanical properties of the polymer, along with a positive influence on the degradation mechanism [39]. However, the presence of transition metals in MMT and their interaction with UV light have also been observed to accelerate the photo-oxidation process [40]. Tcherbi-Narteh [41] also reported that the degradation mechanism varies depending on the clay structure and loading.

Fire Retardants

Flame retardants modify the pyrolysis or oxidation of polymers by slowing or inhibiting the reactions. Development of fire retardants for epoxy has seen accelerating trend in order to extend its applications to various industries, especially electronics and aerospace. There are mainly two ways to impart the fire retardant properties to the epoxy resins. One is the incorporation of the fire retardant additives and the other one involves copolymerization with reactive fire retardants [10]. Halogenated compounds, either as reactive co-reactants or additives, have been widely used as flame retardants traditionally. The release of toxic and corrosive gases during combustion, however, was the major drawback imposing fire hazards and environmental concerns [42]. Thus, the attention nowadays has been focused on the development of halogen-free flame retardants.

Phosphorous derivatives are promising alternative fire retardants for cured epoxy resins, especially 9,10-dihydro-9-oxa-10-phosphaphenanthrene-10-oxide (DOPO) and its derivatives. The efficiency of the phosphorous derivatives originates from their tendency to react with the hydroxyl groups present abundantly in the epoxy structure [10]. The phosphorous structure can also influence the flame retardant performance by affecting the char layer of epoxy [43]. DOPO is mostly used because of its high thermal stability, reactivity and resistance to oxidation [44,45]. Phosphorous derivatives can be applied in epoxy as both additives and reactive co-monomers. The enhanced phosphorous content in the epoxy resin was also observed by Jain *et al.* [46] to improve the limiting oxygen index (LOI). It has been pointed out, however, that organo-phosphorous alone is not capable of conferring the flame retardation properties. The epoxy systems with multiple flame retardants have received attention accordingly.

Yang *et al.* [47] synthesized the flame retardant with phosphorus/nitrogen/boron components and blended it with DGEBA. The results indicated 45.1-72.8% improvement in char yield. Yang *et al.* [48] also conducted copolymerization of DGEBA and phosphorous/nitrogen containing flame retardant (DOPO and N-(4-hydroxyphenyl) maleimide (HPM)). The authors observed improved LOI and decreased total heat release [48]. Boron and silicon containing epoxy also showed a significant reduction in peak heat release rate and total heat release by 69% and 46% respectively as compared to the epoxy resin without flame retardants [49]. With 1.5 wt% boron and 0.5 wt% silicon, the LOI was observed to be 30.5%.

5.3 Conclusions

The thermal decomposition at high temperature with oxidation is observed to be the major pathway for the epoxy resin degradation. The degradation process is generally controlled by oxygen diffusion rather than reaction. Regarding fire retardation, the ratio of carbon to oxygen in the structure is closely linked to the combustion behavior. LOI and char yield are useful indicators to analyze the effect of fire retardants. For stabilization, carbon black has been widely applied for epoxy due to its multitude of properties.

References

1. *Thermal Degradation of Polymer Blends, Composites and Nanocomposites*, Visakh, P. M., and Yoshihiko, A. (eds.), Springer International Publishing, Switzerland (2015).
2. Hamerton, I. (1996) *Recent Developments in Epoxy Resins*, Smithers Rapra, USA.
3. Allen, N. S., and Edge, M. (1992) *Fundamentals of Polymer Degradation and Stabilization*, Elsevier, USA.
4. Minagawa, M. (1989) New developments in polymer stabilization. *Polymer Degradation and Stability,* **25**, 121-141.
5. Saba, N., Tahir, P., and Jawaid, M. (2014) A Review on potentiality of nano filler/natural fiber filled polymer hybrid composites. *Polymers,* **6**(8), 2247-2273.
6. Liu, S., Chevali, V. S., Xu, Z., Hui, D., and Wang, H. (2017) A review of

extending performance of epoxy resins using carbon nanomaterials. *Composites, Part B: Engineering,***136**, 197-214.

7. Christiansen, W., Shirrell, D., Aguirre, B., and Wilkins, J. (2001) Thermal Stability of Electrical Grade Laminates based on Epoxy Resins. Proceedings of *IPC Printed Circuits EXPO*, USA.

8. Mohan, P. (2013) A critical review: The modification, properties, and applications of epoxy resins. *Polymer-Plastics Technology and Engineering*, **52**, 107-125.

9. Levchik, S. V., Camino, G., Luda, M. P., Costa, L., Costes, B., Henry, Y., Morel, E., and Muller, G. (1995) Mechanistic study of thermal behavior and combustion performance of epoxy resins: I homopolymerized TGDDM. *Polymers for Advanced Technologies*, **6**, 53-62.

10. Levchik, S. V., and Weil, E. D. (2004) Thermal decomposition, combustion and flame-retardancy of epoxy resins - A review of the recent literature. *Polymer International*, **53**, 1901-1929.

11. Zahra, Y., Djouani, F., Fayolle, B., Kuntz, M., and Verdu, J. (2014) Thermo-oxidative aging of epoxy coating systems. *Progress in Organic Coatings*, **77**, 380-387.

12. Ernault, E., Richaud, E., and Fayolle, B. (2016) Thermal oxidation of epoxies: Influence of diamine hardener. *Polymer Degradation and Stability*, **134**, 76-86.

13. Gu, A., and Liang, G. (2003) Thermal degradation behaviour and kinetic analysis of epoxy/montmorillonite nanocomposites. *Polymer Degradation and Stability*, **80**, 383-391.

14. Pandey, J. K., Reddy, K. R., Kumar, A. P., and Singh, R. P. (2005) An overview on the degradability of polymer nanocomposites. *Polymer Degradation and Stability*, **88**, 234-250.

15. *Polymer-Clay Nanocomposites*, Pinnavaia, T. J., and Beall, G. W. (eds.), John Wiley & Sons Ltd, UK (2000).

16. Celina, M. C., Dayile, A. R., and Quintana, A. (2013) A perspective on the inherent oxidation sensitivity of epoxy materials. *Polymer*, **54**, 3290-3296.

17. Decelle, J., Huet, N., and Bellenger, V. (2003) Oxidation induced shrinkage for thermally aged epoxy networks. *Polymer Degradation and Stability*, **81**, 239-248.

18. Ernault, E., Richaud, E., and Fayolle, B. (2017) Origin of epoxies embrittlement during oxidative ageing. *Polymer Testing*, **63**, 448-454.

19. Zhang, M., Sun, B., and Gu, B. (2016) Accelerated thermal ageing of epoxy resin and 3-D carbon fiber/epoxy braided composites. *Composites, Part A: Applied Science and Manufacturing*, **85**, 163-171.

20. Zhang, M., Zuo, C., Sun, B., and Gu, B. (2016) Thermal ageing degradation mechanisms on compressive behavior of 3-D braided composites in experimental and numerical study. *Composite Structures*, **140**, 180-191.

21. Yang, Y., Xian, G., Li, H., and Sui, L. (2015) Thermal aging of an anhy-

dride-cured epoxy resin. *Polymer Degradation and Stability,* **118**, 111-119.

22. Galant, C., Fayolle, B., Kuntz, M., and Verdu, J. (2010) Thermal and radio-oxidation of epoxy coatings. *Progress in Organic Coatings,* **69**, 322-329.

23. Djouani, F., Zahra, Y., Fayolle, B., Kuntz, M., and Verdu, J. (2013) Degradation of epoxy coatings under gamma irradiation. *Radiation Physics and Chemistry,* **82**, 54-62.

24. Queiroz, D. P. R., Fraïsse, F., Fayolle, B., Kuntz, M., and Verdu, J. (2010) Radiochemical ageing of epoxy coating for nuclear plants. *Radiation Physics and Chemistry,* **79**, 362-364.

25. Paterson-Jones, J. C., Percy, V. A., Giles, R. G. F., and Stephen, A. M. (1973) The thermal degradation of model compounds of amine-cured epoxide resins. II. The thermal degradation of 1,3-diphenoxypropan-2-ol and 1,3-diphenoxypropene. *Journal of Applied Polymer Science,* **17**, 1877-1887.

26. Luda, M. P., Camino, G., Balabanovich, A., I. and Hornung, A. (2002) Scavenging of halogen in recycling of halogen-based polymer materials. *Macromolecular Symposia,* **180**, 141-151, 3.

27. Liu, Y.-L., Hsiue, G.-H., Lan, C.-W., and Chiu, Y.-S. (1997) Phosphorus-containing epoxy for flame retardance: IV. Kinetics and mechanism of thermal degradation. *Polymer Degradation and Stability,* **56**, 291-299.

28. Rosu, D., Cascaval, C. N., and Rosu, L. (2006) Effect of UV radiation on photolysis of epoxy maleate of bisphenol A. *Journal of Photochemistry and Photobiology A: Chemistry,* **177**, 218-224.

29. Decker, C., Keller, L., Zahouily, K., and Benfarhi, S. (2005) Synthesis of nanocomposite polymers by UV-radiation curing. *Polymer,* **46**, 6640-6648.

30. Liu, F., Yin, M., Xiong, B., Zheng, F., Mao, W., Chen, Z., He, C., Zhao, X., and Fang, P. (2014) Evolution of microstructure of epoxy coating during UV degradation progress studied by slow positron annihilation spectroscopy and electrochemical impedance spectroscopy. *Electrochimica Acta,* **133**, 283-293.

31. Modjinou, T., Versace, D.-L., Abbad-Andaloussi, S., Langlois, V., and Renard, E. (2017) Antibacterial and antioxidant photoinitiated epoxy co-networks of resorcinol and eugenol derivatives. *Materials Today Communications,* **12**, 19-28.

32. Burton, B. L. (1993) The thermooxidative stability of cured epoxy resins. I. *Journal of Applied Polymer Science,* **47**, 1821-1837.

33. Ghasemi-Kahrizsangi, A., Shariatpanahi, H., Neshati, J., and Akbarinezhad, E. (2015) Corrosion behavior of modified nano carbon black/epoxy coating in accelerated conditions. *Applied Surface Science,* **331**, 115-126.

34. Wei, Y. H., Zhang, L. X., and Ke, W. (2007) Evaluation of corrosion

protection of carbon black filled fusion-bonded epoxy coatings on mild steel during exposure to a quiescent 3% NaCl solution. *Corrosion Science,* **49**, 287-302.

35. Neisiany, R. E., Khorasani, S. N., Naeimirad, M., Lee, J. K. Y., and Ramakrishna, S. (2017) Improving mechanical properties of carbon/epoxy composite by incorporating functionalized electrospun polyacrylonitrile nanofibers. *Macromolecular Materials and Engineering,* **302**(5), 1600551.

36. Liu M., and Horrocks, A. R. (2002) Effect of carbon black on UV stability of LLDPE films under artificial weathering conditions. *Polymer Degradation and Stability,* **75**, 485-499.

37. Ghasemi-Kahrizsangi, A., Neshati, J., Shariatpanahi, H., and Akbarinezhad, E. (2015) Improving the UV degradation resistance of epoxy coatings using modified carbon black nanoparticles. *Progress in Organic Coatings,* **85**, 199-207.

38. Ghasemi-Kahrizsangi, A., Shariatpanahi, H., Neshati, J., and Akbarinezhad, E. (2015) Degradation of modified carbon black/epoxy nanocomposite coatings under ultraviolet exposure. *Applied Surface Science,* **353**, 530-539.

39. Tcherbi-Narteh, A., Hosur, M., Triggs, E., Owuor, P., and Jelaani, S. (2014) Viscoelastic and thermal properties of full and partially cured DGEBA epoxy resin composites modified with montmorillonite nanoclay exposed to UV radiation. *Polymer Degradation and Stability,* **101**, 81-91.

40. Morlat-Therias, S., Mailhot, B., Gonzalez, D., and Gardette, J.-L. (2005) Photooxidation of polypropylene/montmorillonite nanocomposites. 2. Interactions with antioxidants. *Chemistry of Materials,* **17**, 1072-1078.

41. Tcherbi-Narteh, A., Hosur, M., Triggs, E., and Jeelani, S. (2013) Thermal stability and degradation of diglycidyl ether of bisphenol A epoxy modified with different nanoclays exposed to UV radiation. *Polymer Degradation and Stability,* **98**, 759-770.

42. Lu S.-Y., and Hamerton, I. (2002) Recent developments in the chemistry of halogen-free flame retardant polymers. *Progress in Polymer Science,* **27**, 1661-1712.

43. Zhang, W., He, X., Song, T., Jiao, Q., and Yang, R. (2014) The influence of the phosphorus-based flame retardant on the flame retardancy of the epoxy resins. *Polymer Degradation and Stability,* **109**, 209-217.

44. Qian, X., Song, L., Hu, Y., Yuen, R. K. K., Chen, L., Guo, Y., Hong, N., and Jiang, S. (2011) Combustion and Thermal degradation mechanism of a novel intumescent flame retardant for epoxy acrylate containing phosphorus and nitrogen. *Industrial & Engineering Chemistry Research,* **50**, 1881-1892.

45. Wang, X., Hu, Y., Song, L., Xing, W., Lu, H., Lv, P., *and* Jie, G. (2010)

Flame retardancy and thermal degradation mechanism of epoxy resin composites based on a DOPO substituted organophosphorus oligomer. *Polymer,* **51**, 2435-2445.

46. Jain, P., Choudhary, V., and Varma, I. K. (2003) Phosphorylated epoxy resin: Effect of phosphorus content on the properties of laminates. *Journal of Fire Sciences,* **21**, 5-16.

47. Yang, S., Zhang, Q., and Hu, Y. (2016) Synthesis of a novel flame retardant containing phosphorus, nitrogen and boron and its application in flame-retardant epoxy resin. *Polymer Degradation and Stability,* **133**, 358-366.

48. Yang, S., Wang, J., Huo, S., Cheng, L., and Wang, M. (2015) Preparation and flame retardancy of an intumescent flame-retardant epoxy resin system constructed by multiple flame-retardant compositions containing phosphorus and nitrogen heterocycle. *Polymer Degradation and Stability,* **119**, 251-259.

49. Yang, H., Wang, X., Yu, B., Song, L., Hu, Y., and Yuen, R. K. K. (2012) Effect of borates on thermal degradation and flame retardancy of epoxy resins using polyhedral oligomeric silsesquioxane as a curing agent. *Thermochimica Acta,* **535**, 71-78.

6

Polymer Nanocomposites for Electronics Applications

6.1 Introduction

Polymer nanocomposites are promising candidates for utilization in electronics industry for electromagnetic shielding materials, embedded capacitors, actuators, cellular phones and sensors [1-5]. A higher dielectric constant or a suitable conductivity are generally required for these applications. These are generally achieved by the appropriate selection of the matrix/filler materials as well as the processing conditions. In the recent past, conductive and elastic polymer nanocomposites have been specifically examined for the electronics applications because of their remarkable deformation capacity, relative to the conventional inflexible metallic strain-sensor gauges as well as conductors. In order to accomplish a moderate electrical conductivity and elastic limit in flexible polymers, electrically conductive nanomaterials, for example, graphene [6], carbon nanotubes (CNTs) [7,8], carbon nanofibers (CNFs) [9] and metal nanoparticles [10], are commonly added reinforcement materials. The inclusion of various fillers in the polymers regulates the comprehensive electrical characteristics of the composites. As an illustration, the incorporation of zinc oxide (ZnO) nanoparticles within the polymer matrix demonstrates a laser-like performance consequent to the optical pumping phenomenon, and the inclusion of barium titanate (BaTiO$_3$) nanoparticles brings about large capacitance in the system. The semiconductor devices based on the functional polymers, hybrids and composites also offer opportunities to be manufactured utilizing comparative printable technologies where distinct dynamic fillers could be incorporated inside the same functional polymer framework, thereby resulting in advanced systems [11-18].

In this chapter, recent advancements and prospects of polymer nanocomposites for use in the electronics industry are reviewed. Four different types of nanofillers, namely graphene, CNTs, CNFs and ZnO are discussed, along with the properties and application of a large variety of polymer nanocomposites. Due to the nano-size of the

Haleema Saleem and Vikas Mittal, Khalifa University of Science and Technology, Abu Dhabi, UAE

reinforcement particles along with optimized filler dispersion in polymer matrices (and subsequent numerous polymer-filler interfacial contacts), the generated polymer nanocomposites display significantly enhanced properties, relative to the pristine polymers or their conventional composites.

6.2 Electronics Applications of Various Polymer Nanocomposites

The nano-sized carbonaceous fillers, for example, graphene, CNTs and CNFs offer excellent electro-mechanical characteristics and higher surface-to-volume ratio. Their high surface vitality assists the process of surface functionalization [19] and can contribute to uniform dispersion as well as effective polymer-filler bonding in organic polymers [20]. In addition, different inorganic nanoparticles, mainly metal oxides, have also been incorporated in the conducting polymers for achieving the desired properties needed for various electronics applications.

6.2.1 Graphene Based Polymer Nanocomposites

Graphene sheets with single atom thickness were initially generated by the mechanical exfoliation, and the material is considered as the thinnest, strongest and lightest material ever developed [21-23]. Due to single atom thickness, it is possible to prepare different materials by efficiently utilizing graphene as atomic scaffolding from where different materials are engineered. Application of graphene in electronics has very high potential because of its capability to expedite electrons or hole transfer along its 2-D surface and its high surface area of almost 2,630 m^{2}g^{-1}. The major issue is that the higher-quality graphene material is an excellent conductor which does not possess a band gap. Hence, in order to utilize graphene in the generation of prospective nano-electronics devices, a band gap is needed to be engineered into it, for diminishing its electron mobility. Overall, the excellent characteristics of graphene outstrip the silicon based electronics devices.

The graphene based polymer nanocomposites have been utilized in touch panel devices, gate dielectrics and electroluminescent devices, because of their good electrical and thermal conductivity, flexibility, smooth fabrication procedure and low temperature processing conditions [24,25]. These materials also find uses in energy storage, nano-electronics devices, biological as well as chemical sensors, etc. Conducting operation in graphene based polymer nanocomposites is described as the development of a continual and electrically conductive system at the percolation threshold, contingent upon the interaction with polymer and nature of dispersion. Graphene with large aspect ratio effectively diminishes the percolation threshold in polymer nanocomposites, thereby, allowing to prepare superior conducting films at low filler concentration. Also, the greater reduction temperature of graphene oxide (GO) generates enhanced electrical conductivity, relative to increasing the reduction time at lower temperatures. In the energy storage uses, the graphene based polymer nanocomposites outstrip the typical rechargeable lithium batteries due to the fact that these materials have excellent energy and power density. Also, graphene based polymer nanocomposites are regarded as supercapacitors [26,27]. Graphene improves the piezoelectric characteristics of the polymer without influencing the elastic modulus.

In the following sections, the use of graphene based polymer nanocomposites in different applications such as fuels cells, solar cells, sensors, photovoltaics and photodetectors has been reviewed in brief.

Fuel Cells

The fuel cells contribute beneficial transformation of energy because of lower emission of greenhouse gases and greater power density [28-30]. In a study by Wang *et al.* [31], the impact of graphene based polymer nanocomposites in the organic solar devices was studied. The device efficiency was examined for solar cells utilizing the blends consisting of the conjugated polymer, poly(3-hexylthiophene) (P3HT), with two different functionalized electron acceptors: commercially available [6,6]-phenyl C-60 butyric acid methyl ester (PC60BM) as well as graphene derivative for enhancing the carrier mobility. It was noted that the energy conversion efficiency of the device with lower loading of solution-processable functionalized graphene (SPFG) was enhanced remarkably, relative to the device with no SPFG. The enhanced efficiency was attributed to the presence of semiconducting graphene which contributed charge-transportation paths for the carriers to be extracted and consequently increased the carrier mobility in the organic photovoltaic devices. Nevertheless, as the SPFG concentration was enhanced to a certain level, the efficiency of the device diminished. The graphene doped conducting polymers like poly(3-hexylthiophene), poly(3,4-ethylenedioxythiophene) and poly(styrene sulphonate) have also been observed to exhibit superior power consumption effectiveness of 4.5%, relative to pristine polymer counter electrode [32,33]. The functionalized GOs show exclusive phosphate adsorption capability, onset potential, electro-chemiluminescence activity, stability, ion-exchange capacity, H_2O_2 electrochemical detection and good solubility. By utilizing the microwave reduction strategy, these functionalized graphene materials have also been utilized to generate polymer electrolyte membrane fuel cells (PEMFC).

At a 10 V bias voltage, the electrical conductivity of hydrophilic pristine GO is about 5×10^{-3} S.cm^{-1}, and it is considered as an electronic insulator [34]. The GO filled polymer nanocomposites utilizing flexible and high strength poly(ethylene oxide) (PEO) resulted in high protonic/ionic conductivity, due to the presence of ample protons in GO/PEO nanocomposite membrane. At lower temperatures of 20°C - 60 °C, for the fuel cells, greater conductivity caused the generation of low temperature membrane materials. Comparison between Nafion and GO based PVA membranes confirmed the attractive properties like enhanced conductivity as well as diminished biofouling on the membranes, thereby, resulting in higher power densities and improved power generation in microbial fuel cells (MFCs) [35]. The high surface area and exclusive structure of GO contributes additional proton transportation channels and also holds extra water, which can be advantageous for the enhancement of the proton conductivity together with the mechanical characteristics of the generated membranes [36].

It is noted that the change in the loadings of GO and GO functionalization influence the membrane biofouling. Thus, the MFCs with graphene electrochemically deposited on carbon cloth [37] as well as polyaniline (PANI) hybridized three-dimensional (3-D) graphene have been generated [38]. For instance, Yong *et al.* [38] illustrated the application of PANI/3-D graphene framework as the MFC anode. Three dimensional graphene/PANI structure was observed to have superior performance than the frequently utilized carbon cloth because of the higher extracellular electron transfer (EET) efficiency and greater bacterial biofilm loading as presented in Figure 6.1. This is because of the higher specific surface area (SSA) of the PANI/3-D graphene electrode along with its capability to integrate with the bacterial biofilms three-dimensionally.

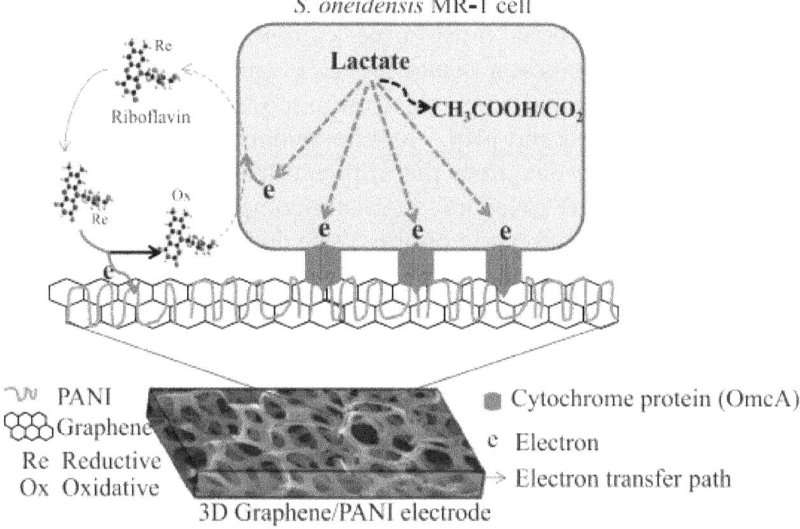

Figure 6.1 Schematic representation of the interface between S. *oneidensis* MR-1 bacteria and PANI/3-D graphene monolith electrode. Reproduced from Reference 38 with permission from American Chemical Society.

Photovoltaics and Photodetectors

In the photovoltaic cells (PV cells), graphene nanofiller has been utilized in many ways - transparent counter electrodes, electron acceptor materials, hole/electron transportation and electron acceptor materials, buffer or interfacial layers for retarding the charge recombination inside the components, etc. Nevertheless, large scale preparation of graphene in its sheet form and uniform dispersion in the polymer materials are the major challenges. For bulk heterojunction polymer organic PV cells, the graphene based materials are utilized as window electrodes with simple processing and flexible substrate compatibility [39]. The solar energy transformation to electricity mostly happens by the process of light absorption exciton production, diffusion of exciton, separation as well as charge generation, charge transportation and accumulation. In the conjugated polymers, detachment of the photo-excited electron-hole pair is accomplished by generating a heterojunction possessing an acceptor material having greater affinity for the electrons, relative to the polymer. The aforementioned heterojunction contributes the driving force for dissociating the firmly bound electron-hole pairs into distinct charges. Acceptor material must possess a moderately lesser HOMO (highest occupied molecular orbital) level, as compared to the conjugated polymer, for transferring electron to it and to keep the hole in the polymer valence band [40]. The effectiveness of the planar heterojunction device relies upon the diffusion length of excitons, and the different graphene based polymer nanocomposites fulfill the fundamental criteria for adequate photovoltaic devices [41].

A poly(3-octyl-thiophene)(P3OT)/graphene nanocomposite based unique photo-electrochemical cell (PEC) was developed by Chang *et al.* [42] for the photovoltaic solar energy transformation. Owing to the non-covalent functionalization using pyrenebutyrate (PB), solution processable graphene (SPG) was accomplished and employed to generate the P3OT/graphene nanocomposites for PEC utilization. It was demonstrated that the graphene doping in the P3OT film remarkably enhanced the photocurrent and photovoltaic conversion effectiveness of PEC cells by almost 10 folds. Also, the efficiency of the photo-electrochemical cells was to a great extent reliant on the morphology of the P3OT/graphene nanocomposites and graphene content. The maximum photovoltaic conversion effectiveness was acquired at 5 wt% graphene concentration in the nanocomposites. The SPG/P3OT nanocomposite PEC contributes a common platform for next generation photodetectors, photoconductivity and solar energy conversion. Hsu *et al.* [39] also reported a molecular doping procedure in a layer-by-layer manner on graphene for generating sandwiched graphene/tetracyanoquinodimethane (TCNQ) stacked films for polymer based solar cells. Figure 6.2 shows the schematic of the synthesis process.

The performance of the solar cell relies upon the inherent electronic properties of the distinctive materials, preparation conditions and structure of the device [43]. Alternate structural characteristics such as polymer purity, polydispersity, molecular weight and regioregularity additionally influence the performance of solar cells. The pure graphene shows secondary device efficiency because of its low compatibility with organic polymers. Due to the presence of chemically reactive oxygen functionality at GO basal planes, the utilization of GO is beneficial, relative to graphene. Surface and chemical modifications of graphene could additionally be performed to enhance the surface characteristics as well as electrical conductivity of graphene, thus, advancing the polymer-graphene electronic interactions for optimal performance.

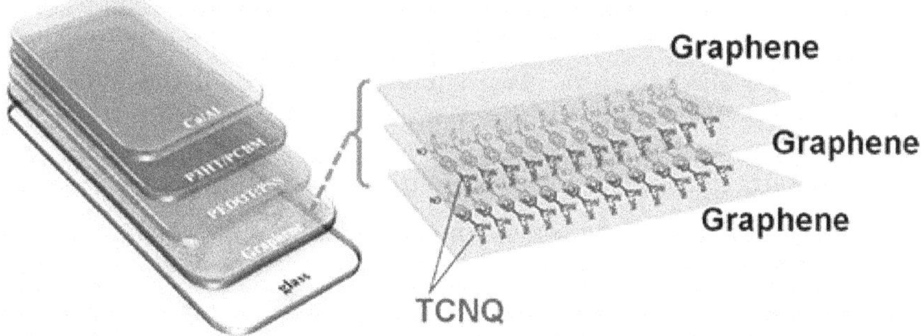

Figure 6.2 Schematics of the molecular doping process. Reproduced from Reference 39 with permission from American Chemical Society.

Sensors

In the recent past, there is an enhanced interest in the field of strain detection in structural health monitoring (SHM), utilizing unique polymer nanocomposite systems. Eswaraiah *et al.* [44] prepared functionalized graphene (f-Graphene)-polyvinylidene fluoride (PVDF) using shear combined with solvent casting method. PVDF was chosen due to its novel properties like temperature sustainability, chemical resistance and its utilization in pyroelectric and piezoelectric fields, among others. Uniform as well as stable dispersion of graphene in the polymer solution was accomplished subsequent to the functionalization. The PVDF/graphene nanocomposites exhibited 3-D interconnecting conducting framework of graphene within the polymer. The investigation of the variation in voltages of various nanocomposite films confirmed that the PVDF nanocomposite film with 2 wt% f-Graphene had improved strain detection performance, relative to its carbon nanotube polymer composite counterpart. The quick variation in the tunneling and contact resistance in the polymer caused a better efficiency in the sensing of strain, which could be attributed to the adaptability of graphene at optimum concentration.

The conjugated polymers, for example, polythiophene, polypyrrole and PANI have been broadly analyzed for the development of gas sensors, depending on their sensitivities, ambient temperature utilization as well as response times [45]. The electronic framework of these polymers consists of conjugated π-bonds, which could experience variation under the impact of the chemical species adsorbed on the surface, because of an acid-base type or redox type interaction between the chemical species and polymer. The preparation as well as device utilization of a PANI/graphene nanocomposite was reported by Al-Mashat *et al.* [46]. The authors studied the hydrogen (H_2) gas detection ability of the nanocomposite and contrasted the performance with the sensors based on PANI nanofibers and graphene independently. It was noted that the graphene based polyaniline composite-based gas sensor possessed sensitivity of 16.57% towards 1% of H_2 gas, which was higher as compared to the sensitivities of the sensors based on polyaniline nanofibers (9.38%) and graphene (0.83%) individually. It was also noted (Figure 6.3) that for lower concentrations of H_2 gas, the sensitivities of the polyaniline nanofibers and polyaniline/graphene nanocomposite-based sensors were equivalent. This was due to the fact that similar doping level of PANI was accomplished for the two sensors.

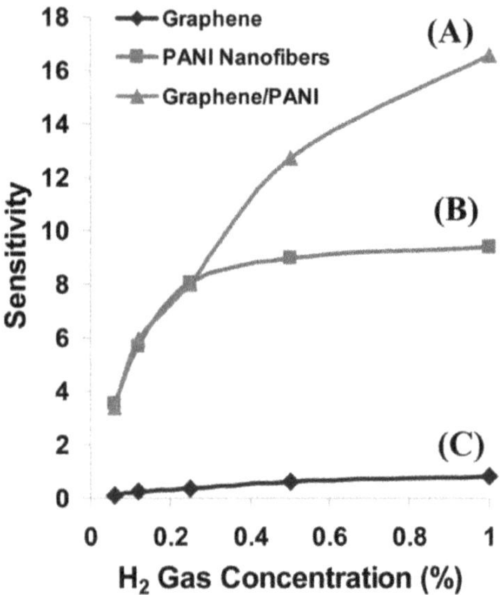

Figure 6.3 The hydrogen sensitivities of gas sensors with sensitive layers of (A) polyaniline/graphene, (B) nanofibers of polyaniline and (C) graphene. Reproduced from Reference 46 with permission from American Chemical Society.

For completely transferring the superlative characteristics of graphene into a polymer and generate a high performance nanocomposite, certain factors, for example, orientation, preparation strategy, interface between nanofiller and polymer as well as optimal dispersion of nanofiller, are involved. Several efforts have been made to prepare the single layer graphene and its uniform dispersion in polymer frameworks, which is fundamental for the generation of electrically conducting graphene nanocomposites. Eswaraiah *et al.* [47] reported a one-step strategy for the generation of conducting reduced graphite oxide/PVDF nanocomposites, utilizing PVDF and graphite oxide (GO) as starting materials. The preparation was achieved completely through a dry route. GO was observed to be totally reduced in PVDF in the dry condition, utilizing the photochemical reduction. The synchronous decrease and ultra-quick melting of the polymer prevented the graphene restacking. The generated composites were observed to be remarkably conducting and displayed lower percolation thresholds. A strain gauge factor of 12.1 was attained, thus, confirming the promising use of the PVDF/r-GO nanocomposite as a potential strain sensor in SHM.

6.2.2 CNTs Based Polymer Nanocomposites

The carbon nanotubes comprise of at least one sheet of graphene rolled in cylindrical form. Independently, the nanotubes can be semiconducting or metallic relying upon their diameter and chirality [48]. The CNTs have been widely examined as they exhibit 0.27-1.25 TPa Young's modulus, 10^4-10^5 S/cm electrical conductivity, 3000-6600 W/m.K thermal conductivity, 1000-4000 cm^2/V.s carrier mobility and thermal stability till 700 °C in air [49,50]. These electronic, thermal and mechanical properties enable CNTs to be used as reinforcing materials for generating superior multi-functional polymer nanocomposites [51]. Specifically, the electrically conductive polymer nanocomposites with CNTs exhibit the potential of applications in advanced areas, for example, supercapacitors [52,53], electromagnetic shielding [54,55], smart actuators [56,57], energy harvesting [58], sensors [59,60] and different biomedical devices [61,62]. For the single-walled carbon nanotubes (SWCNTs), the lengths are up to 1.5 cm, whereas the diameters are in 0.4 to 2 nm range [63]. These dimensions bring about aspect ratios of about ten million, which is to a greater degree favorable for achieving much lower percolation thresholds in polymer composites. The multi-walled carbon nanotubes (MWCNTs) consists of concentric graphene layers which are spaced 0.34 nm apart, having lengths up to a few microns and diameters from two to a few hundred nanometers. One specific kind of MWCNTs are the double-walled carbon nanotubes (DWCNTs), generated by two concentric graphene sheets. As the polymer/CNT nanocomposites exhibit features like cost-effectiveness, scalability, mechanical flexibility, lightweight nature, easy processability and compliance, these materials can also be utilized as the electric heating elements [64]. Mostly, these electric heating devices or materials, which are a type of electrical resistors which change the electrical energy to thermal energy by colliding the mobile electrons with the atoms which make up the conductor body, are utilized for a variety of applications, including medical instruments, industrial processes, etc. In the following section, the potential applications of CNT based polymer nanocomposites in supercapacitors, electromagnetic shielding, sensors, photoconductive and photovoltaic devices have been briefly summarized.

Supercapacitors

For supercapacitors, CNTs represent a functional reinforcement for the conducting polymers. The exclusive conducting, mechanical and micro-textural properties of CNTs permit to generate a supercapacitor possessing an appreciable electrode/electrolyte interface for effective propagation of charge. The electrically conducting polymers (ECPs) serve as an exceptionally intriguing family of synthetic metals because of a quick electrochemical switching and significant doping level. The use of ECPs in the electrochemical capacitors is directed by their remarkable capacitance values [65,66]. The primary disadvantage of the utilization of ECP as the supercapacitor electrodes is associated with their inadequate stability at the time of cycling. The ECP films, because of the volumetric variation throughout the course of dedoping/doping procedure, experience shrinkage, swelling, breaking or cracking, which moderately disturbs their conducting characteristics. Furthermore, the electrochemical activeness of all ECPs is entirely controlled by their working capability range, restricted by a secluding state or/and degradation of polymer induced by the over-oxidation process. Generation of their composites with CNTs helps to overcome these limitations to a large extent.

Frackowiak *et al.* [52] attempted to resolve the ECP stability issue with cycling process by utilizing the composites with CNTs as the mesoporous conducting system capable of adjusting to the entire mechanical stress. CNT composites of three types of ECPs, i.e., poly-(3,4-ethylenedioxythiophene) (PEDOT), polypyrrole (PPy) and PANI, were examined as the supercapacitor electrode materials. Long-term charge/discharge tests at current load of 300 mA/g demonstrated a satisfactory cycleability of CNT/ECP composites. The impedance spectroscopy proved better accessibility of ions within ECP subsequent to 3000 cycles exclusively for composites having CNTs, while the pristine PANI displayed an outstanding resistive increment. Applied potential was noted to be the major aspect affecting the nanocomposites with conducting polymers. In order to operate all electrodes in optimum potential range, the unsymmetrical capacitors were generated with MWCNT/PANI and MWCNT/PPy as positive and negative electrodes, respectively. Greater capacitance values till 320 F/g of electrode material were attained with this arrangement. A supplementary development of capacitor voltage was achieved utilizing the CNT/ECP composite as the positive electrode and the activated carbon as the negative electrode.

Electromagnetic Shielding

Due to their versatility, processability and lightweight, the conducting polymer composites (CPCs) are potential candidates for the application in different electrical and electronic devices in order to fulfill the electromagnetic compatibility prerequisites. A device is said to be electromagnetically adaptable with its surrounding in the event that it does not intervene with itself or alternate devices, and it is not influenced by the emanations from additional devices [67,68]. Accordingly, an adequate shielding material must avoid both outgoing and incoming electromagnetic interference (EMI). The shielding effectiveness (SE) of EMI is indicated in terms of decibel (dB). A SE of 30 dB, relating to 99.9% EMI radiation attenuation, is regarded to be a sufficient shielding level for several applications [69]. For the EMI shielding, metal plated or metal coated polymers are the most broadly utilized materials [70]. Further, typical CPCs made of carbon fibers, stainless steel fibers and Ni coated carbon fibers have likewise been utilized as EMI protecting enclosures, however, to a lower degree [71], due to the higher filler concentration needed to accomplish a satisfactory degree of shielding. Higher filler concentration increases the cost of composites, thereby, restricting the commercial application. After the development of CNTs and CNFs as reinforcing fillers, various studies were also reported on the EMI shielding effectiveness of CNFs as well as CNTs based polymer nanocomposites [72-75]. The experimental outcomes illustrated that at a similar filler concentration, polymers reinforced using the nanosized carbon filler possessed greater EMI shielding effectiveness, relative to the polymer reinforced using nanosized carbon black (HS-CB) or micro-scale carbon filler.

Designing an economically as well as technically competitive CPC for meeting the electromagnetic compatibility standards is very important due to the fact that over-shielding enhances the weight, cost and processing complications of composite, whereas the under-shielding might bring about the failure of the product [67]. Proper comprehension of the CPC EMI shielding mechanism is critical for the foremost utilization of the CPC EMI shielding abilities and for designing a conductive polymer composite at the optimal filler loading as well as cost. The electromagnetic interference shielding mechanisms of MWCNT based polymer nanocomposites have been analyzed experimentally as well as hypothetically by Al-Saleh and Sundararaj [54]. For the experimental investigation, the EMI shielding effectiveness of MWCNT based polypropylene (PP) nanocomposite plates, generated at four distinct concentrations and in three distinct thicknesses, was studied. In the EMI shielding, three mechanisms namely reflection, multiple-reflection and absorption are involved. In the case of a homogeneous conductive material sheet, reflection is generally considered to be the fundamental shielding procedure [76]. For shielding by the reflection procedure, the material should possess portable charge carriers for interacting with the approaching EM waves. Second critical procedure is absorption, which relies upon the shield thickness. Multiple-reflection is the third type of shielding procedure, which illustrates the reflections inside the shielding material. For polymer composites, the shielding procedures are more convoluted relative to homogeneous conductive obstructions, due to the higher surface area accessible for multiple-reflection as well as reflection phenomena. The hypothetical examination results demonstrated that the multiple-reflection inside the CNT inner surfaces may negatively impact the overall EMI shielding effectiveness. It was also noted that the multiple-reflection among the CNT outside surfaces additionally diminishes the overall shielding, however, its impact is smaller than that observed inside the inner surfaces. The experimental outcomes illustrated that the shielding by means of absorption for MWCNT based PP nanocomposites relies upon the separation between the CNT particles or/and the nanocomposite's electrical resistivity. The shielding by reflection mechanism was noted to rely upon the thickness and conductivity of the nanocomposite specimen and the MWCNT concentration. Utilization of a multi-surface shield was opined to essentially improve the overall EMI shielding effectiveness provided that the multiple-reflection is limited.

Lee *et al.* [55] reported a novel strategy for improving the dispersion of modified MWCNTs in PP matrix for the electrostatic discharge (ESD) application. The MWCNTs were surface functionalized with octadecylamine (ODA) by means of CF_4 plasma-supported fluorination and consequent alkyl-amination. The amount of fluorine groups on the surface of MWCNTs was regulated by changing the CF_4 plasma treatment settings. The XPS and Fourier transform infrared spectroscopy results confirmed that the reaction between ODA and fluorinated MWCNTs was unequivocally influenced by the reaction temperature and solvent quality. The resultant MWCNTs based PP nanocomposites exhibited a substantially better dispersion in the polymer matrix, relative to the non-modified MWCNTs, thereby, resulting in an enhanced electrical conductance at lower MWCNTs concentration (2 wt%). Figure 6.4 presents the surface resistivity of PP nanocomposites filled with p-MWCNTs or ODA-MWCNTs as a function of CNTs

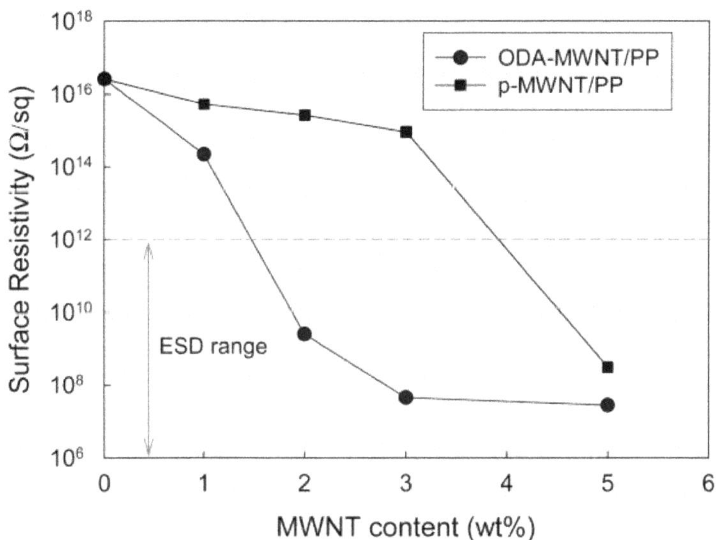

Figure 6.4 Surface resistivity of PP nanocomposites filled with p-MWCNTs or ODA-MWCNTs as a function of the CNTs concentration. Reproduced from Reference 55 with permission from American Chemical Society.

concentration. An intense reduction in the surface resistivity was noted for 5 wt% p-MWCNTs nanocomposites and 2 wt% ODA-

MWCNTs nanocomposites. The stepwise variation in the surface resistivity could be considered as an electrical percolation threshold. As the CNTs concentration reaches a conductivity threshold, the conductive nanotubes framework occurs in the nanocomposite, thereby, generating a conductive pathway. In addition to this, the PP nanocomposites demonstrated remarkably enhanced mechanical properties, and the complex viscosity as well as storage modulus enhanced considerably in the low frequency zone as the CNTs concentration was enhanced, demonstrating a rheological percolation threshold at 1 wt% CNTs concentration.

Sensors

Electrically conductive polymer composites (ECPCs), comprising of an insulating polymer and a conductive filler, have acquired significant consideration because of their novel electrical and mechanical properties [77,78]. Research studies have demonstrated that the resistance of ECPCs relies upon the outside pressure, an event termed as the piezo-resistive effect [7,79]. The mechanical flexibility, processing simplicity and piezo-resistivity of the ECPCs enable these composites as potential materials for the applications in flexible, inexpensive and high area pressure sensors [80,81]. The ECPC's conduction process can be demonstrated with reference to a conductive filler system in a polymer, where the critical concentration (C_c) of the filler for the network development is termed as the percolation threshold [82]. The exceptionally greater aspect ratio of CNTs implies that these can generate a conductive system in a polymer matrix at a much lower percolation threshold, relative to alternate conductive fillers like metallic nanoparticles and carbon black, which have much lower aspect ratios [83,84]. Also, the ECPC's percolation threshold is affected by several factors like the polymer matrix viscosity, shape of filler particles and degree of polymer-filler interactions [7]. Among the aforementioned factors, accomplishing a homogeneous distribution of the conductive filler within the polymer matrix is pivotal for bringing down the percolation threshold.

Hwang *et al.* [59] manufactured a piezo-resistive composite utilizing polydimethylsiloxane (PDMS) as the polymer network and MWCNTs as the conductive filler, which conducted in a remarkably low pressure range needed for finger-sensing. In order to accomplish a uniform dispersion of the CNTs in PDMS, the CNTs were modified utilizing poly(3-hexylthiophene) (P3HT), by a polymer wrapping strategy. The percolation threshold of the nanocomposites was re-markably diminished by the presence of P3HT. The piezo-resistive sensitivity and electrical conductivity of the nanocomposite were also noted to rely upon the concentration of P3HT. As a result, the PDMS/P3HT-MWCNT nanocomposite demonstrated enhanced pi-ezo-resistive property in the pressure range 0-0.12 MPa.

The piezo-resistive characteristics of conductive framework of CNTs in the polymer matrices are accountable for the strain detec-tion. The nanocomposite film sensor's electrical resistance can be contingent on different tensile strains for determining the sensor sensitivity. As the CNT network is changed by the mechanical defor-mation, the resistivity enhances. The aforestated resistivity variation is related to the modification of contact arrangements as well as tun-neling separation between CNTs. The utilization of SWCNTs and polymethylmethacrylate (PMMA)/SWCNTs nanocomposites has been reported for the sensing of strain [85]. Strain response of SWCNTs is greater, however, they are not suitable for analyzing the strain in the entire elastic range. In the PMMA/SWCNTs nanocompo-sites, the slippage is prevented due to the bonding between the CNTs and polymer, thereby, adequately enhancing the sensor strain. Pham *et al.* [86] examined the resistivity of MWCNTs based polymer nano-composite films under tensile strain and demonstrated their poten-tial utilization as stain sensors with customized sensitivity. Wang *et al.* [87] investigated the piezo-resistivity of MWCNTs filled silicone rubber nanocomposites under the influence of pressure. The active carboxyl radicals on the CNTs could improve the uniform distribution as well as the positioning of the conductive pathways in the nanocom-posites. Subsequently, the nanocomposites demonstrated positive pi-ezo-resistance with enhanced sensitivity together with linearity for pressure. Overall, due to the greater sensitivity at room temperature, the CNT based gas sensors have obtained considerable attention. The electrical conductivity varies with adsorption of different gases [88].

The opto-electronic materials, which are sensitive at the wave-lengths in the near-infrared (NIR) domain (for example, 800-2000 nm), are advantageous for challenging applications such as remote sensing, solar cells, telecommunication, thermal photovoltaics and thermal imaging [89], especially when incorporated with CNTs, which contribute to the feasibility of a consolidated electronics as well as opto-electronics technology [90]. Pradhan *et al.* [91] reported that the infrared (IR) photo-response in the electrical conductivity was drastically improved by incorporating SWCNTs in a thermally and electrically insulating polymer. Figure 6.5 presents the schematic of device framework. The scanning electron micrograph demonstrated uniform dispersion of poly(p-phenylene ethynylene)-SWCNT$_{HiPco}$ in polycarbonate, which was necessary for generating the composites exhibiting isotropic electrical conductivity. Conductivity variation in a 5 wt% SWCNT-polycarbonate nanocomposite was remarkable (4.26%) as well as sharp consequent to IR illumination, at room temperature, in air. Even though the thermal impact dominated in the IR photo-response of a pristine SWCNT film, the photo-effect dominated in the IR photo-response of SWCNT nanocomposites.

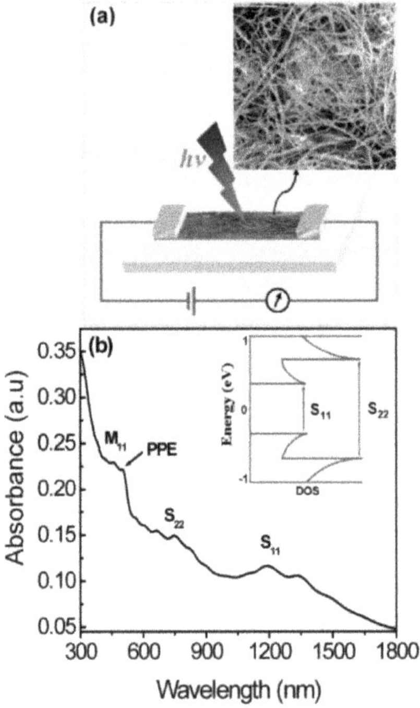

Figure 6.5 Schematic of device framework. Inset: micrograph of the 5 wt% SWCNT$_{HiPco}$-polycarbonate nanocomposite film. Reproduced from Reference 91 with permission from American Chemical Society.

Photoconductive and Photovoltaic Devices

The electrical, optical and structural properties of SWCNTs led to their investigation for utilization in the NIR photoconductive (PC) as well as photovoltaic (PV) devices. Kazaoui *et al.* [58] generated the prototypical Al/SWCNT-polymer/indium tin oxide (ITO) thin-film devices demonstrating potential PV as well as PC responses from ultra violet (UV)-visible to NIR, mostly in the 300-1600 nm spectrum. The observed performance could be attained by uniformly dispersing the SWCNT powder within the polymers like polythiophene and poly(phenylene-vinylene). The authors observed that the devices used (1) the electronic transportation characteristics of the polymer matrices in consolidation with that of the SWCNTs, (2) the NIR light harvesting characteristics of semiconducting SWCNTs and (3) the energy/charge transfer processes between the polymers and the CNTs. The authors also revealed that the characteristics of SWCNTs based polymer thin-film devices are tunable. It was also found that the variations of short-circuit current (SSC) were mostly associated with the light harvesting characteristics of semiconducting SWCNTs. Counter to PbS, C_{60} or ZnSe/InAs nanoparticles in polymer films, the authors observed that the semiconducting SWCNTs have higher potential to enhance the effectiveness of photovoltaic cells and photodetectors.

The advancement of semiconducting polymers as an advanced class of opto-electronic materials has generated the interest with respect to their utilization in the field of photovoltaic cells [58] and photodetectors [92]. The photovoltaic characteristics of blend composites based on dye (N-(1-pyrenyl)maleimide (PM))] functionalized SWCNT-conjugated polymer (P3OT) were reported by Bhattacharyya *et al.* [93]. The authors utilized P3OT as the photo-excited electron donor, which was mixed with PM functionalized SWCNTs as well as with pristine SWCNTs to generate two classes of photovoltaic cells namely ITO/P3OT-SWCNT+PM/Al and ITO/P3OT-SWCNTs/Al, respectively. Figure 6.6(a) represents the non-covalent functionalization of the CNTs with PM molecules. Photocells were prepared in the sandwich configuration, as presented in Figure 6.6(b). Enhanced performance was observed owing to the functionalization of SWCNT using the dye molecules. The authors noted that the short circuit current (SCC) was enhanced by more than one order of magnitude, relative to the SWCNT-polymer diode with no dye. The principle reason for the development in the SCC was attributed as the proficient hole transfer by the dye molecules to poly(3-octylthiophene) at polymer/dye interface and quick transmission of generated electrons to the SWCNTs at the interface between CNTs and dye. The developed materials represent a functional class of organic semiconducting materials with a variety of potential applications.

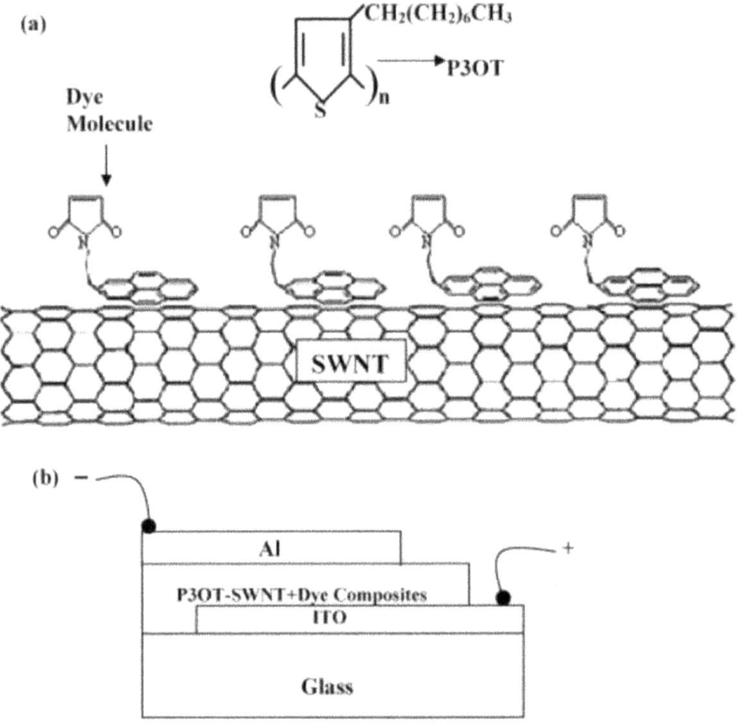

Figure 6.6 Chemical structure of P3OT and PM, along with schematic of PM attachment to the surface of SWCNT by means of π-stacking; (b) device framework of photovoltaic cell. Reproduced from Reference 93 with permission from American Chemical Society.

The organic solar cells are potential alternatives to the silicon solar cells, however, the fundamental disadvantage of these devices is low power conversion effectiveness. The cells generated using two electrodes and a solitary polymer have the tendency to be ineffective due to the fact that the photo-generated excitons are normally not split by the generated electric field because of the contrasts in the electrode work capacities. Additionally, when the free carriers are generated, these possess low mobilities and the recombination procedure amongst holes and electrons is exceptionally ideal. A technique for resolving the aforestated drawback is the incorporation of CNTs to generate an interpenetrating mix with the polymer. Subsequent to the discovery of photo-induced charge transmission between CNTs and organic conjugated polymers, the CNTs are utilized to manufacture PV devices in consolidation with various polymers for increasing the power conversion effectiveness. Impact of various types of CNTs like DWCNTs, SWCNTs and MWCNTs in organic solar cells was analyzed by Arranz-Andres and Blau [94]. The CNT incorporation in P3HT matrix was observed to change the polymer framework, and, thereby, device properties. The device's working principle is that the CNT interaction with the polymer brings about local order of polymer chains. In addition, these interactions permit charge segregation of the photo-generated excitons in the polymer and productive transportation of the electrons to the cathode by way of CNTs. This explains for the improvement in performance of the photovoltaic devices. Enhanced charge detachment as well as collection was illustrated by the large variation between dark and light conductivities, and additionally the increment in SSC, fill factor and open-circuit voltage (OCV). Energy level modification of poly(3-hexiltiophene) with the incorporation of CNTs caused greater OCV. For most of the CNTs analyzed, the CNT C_c value of 5 wt% was noted to provide the most noteworthy OCV. The CNT incorporation to the polymer enhanced the power transformation effectiveness by 3 orders of magnitude, relative to the device with no CNTs. The results demonstrated that the CNT-conjugated polymer nanocomposites depict an alternate approach for generating organic PV cells possessing enhanced efficiency. Additional advancements in the performance of device can be anticipated with more controlled film generation as well as polymer doping.

6.2.3 CNFs Based Polymer Nanocomposites

In most cases, CNFs have larger diameters and lower crystallinity as compared to CNTs. These materials are hollow core nanofibers possessing greater aspect ratios and including a distinct layer, or a double layer of graphite planes stacked parallel at a specific angle to the axis of fiber [95]. Contingent upon the feedstock, operating conditions and catalyst, CNFs with various morphologies and physical attributes can be acquired. In this section, the application of CNFs based polymer nanocomposites in electromagnetic shielding and sensors has been briefly reviewed.

Electromagnetic Shielding

As mentioned earlier, the electromagnetic interference persists as a technical challenge with respect to the fast development of electronic devices, for example, medical instruments, computers and wireless communication devices, which run at greater frequencies in tinier packages. The electronic and electrical components should be shielded for the entire range of the electromagnetic interference frequency spectrum. Carbon fibers are generally very effective for the shielding application [96], whereas majority polymers are EMI transparent. The vapor grown carbon nanofibers (VGCNFs) are considered to be effective nanofillers for the polymers for EMI shielding due to their high modulus, strength, surface area and electrical as well as thermal properties. Several studies have concentrated on the investigation of anisotropic and isotropic VGCNF polymer composites and have demonstrated improved electrical, mechanical as well as thermal properties and better EMI shielding effect [97,98].

VGCNF filled liquid crystal polymer (LCP) composites were examined by Yang *et al.* [75]. LCP resins possess good thermal and flame stability, mechanical properties and chemical resistance. These are particularly suitable to generate homogenous molecular orientation under the elongational flow. Such resins are generally utilized in electronic and electrical components, connectors and cables for fiber optics, devices for chemicals processes, automotive and mechanical engineering, medical equipment and aerospace industry. The percolation theory demonstrates that the composite's electrical conductivity is contingent upon the capability to generate a conducting system [98]. Thus, due to long rigid rod molecules and profoundly ordered fluid condition, it is anticipated that the LCP resins will support the development of a conducting system with a lower nanofiber concentration. Yang *et al.* [75] observed that the surface resistivity of LCP diminished more than 11 orders of magnitude by adding 5 wt% of CNFs. The electromagnetic interference shielding effectiveness of the nanocomposites at various frequencies was additionally contemplated. It was found that the composites displayed till 41 dB of SE. The system's shielding mechanism was fundamentally reflection as well as multiple-reflection. In addition, the thermal conductivity of the composites demonstrated no remarkable advancement with the incorporation of CNFs.

Sensors

Flexible as well as deformable electrically conductive materials are beneficial for developing wearable electronics. For addressing the challenge of developing flexible materials with a moderately high electrical conductivity and advanced elastic limit, Wu *et al.* [99] reported a simple technique to generate porous PDMS(p-PDMS)/CNF nanocomposites (Figure 6.7), utilizing CNF coated sugar particles as the templates. The resultant 3-D porous nanocomposites, with CNFs incorporated in the pore walls of PDMS, demonstrated a remarkably enhanced failure strain (till almost 94%), relative to the solid pristine PDMS (almost 48%). Piezo-resistive behavior under the cyclic tension illustrated that the novel microstructure contributed superior durability to the nanocomposites. The gauge factor and electrical conductivity of the nanocomposites could be modified by varying the CNFs content. The nanocomposites demonstrated stable piezo-resistive behavior with quick response time and linearity in the $\ln(R/R_0)$ vs. $\ln(L/L_0)$ till approximately 70% strain. The aforementioned adjustable sensitivity as well as conductivity confirmed the potential of these stretchable nanocomposites for the utilization as flexible strain-sensors for monitoring the movement of human joints and additionally as flexible conductors for wearable electronics.

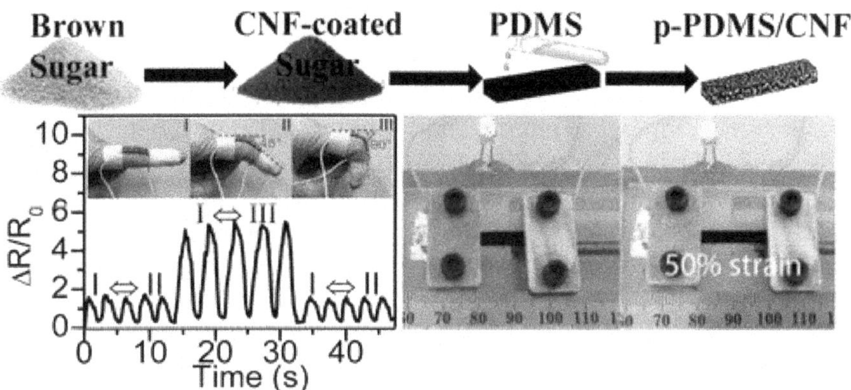

Figure 6.7 Schematic of the synthesis as well as performance of the polymer/CNF composites. Reproduced from reference 99 with permission from American Chemical Society.

Polymer nanocomposites consisting of a plasticized thermoplastic or a crosslinked elastomer and CNFs at concentrations higher than the percolation threshold are noted to demonstrate a distinctively strain-reversible piezo-resistive behavior consequent to the implementation of quasi-static tensile strain [9]. Toprakci *et al.* [9] noted that at lower strain levels, the electrical resistance of the nanocomposite decreases with enhancing strain, which is considered to be an indication of negative piezo-resistivity. Beyond a critical strain, nonetheless, with enhancing strain, the resistance reverses as well as enhances, indicating the presence of a negative-to-positive piezo-resistivity progress, which is completely strain-reversible and repeatable subsequent to the strain cycling. These features suggested stable morphologies of the nanocomposites with little mechanical hysteresis. Process underlying this progress is attributed to the reorientation of high aspect ratio CNFs at lower strains, followed by separation at higher strains. Deposition of these nanocomposites as print coatings on the textile fabric maintains the strain-reversible piezo-resistivity, revealing the fact that these materials are appropriate as printable strain sensors.

6.2.4 ZnO Based Polymer Nanocomposites

Out of the different classes of inorganic nanoparticles, zinc oxide (ZnO) has attained consideration research attention due to its remarkable catalytic, electronic, optical and electrical properties [100]. ZnO is a polar group II–VI semiconductor material having a direct bandgap of almost 3.37 eV. It also possesses several benefits over other wide-bandgap semiconductors. ZnO nanoparticles have been extensively employed in paints and rubber processing, whereas polycrystalline forms of ZnO have been utilized for several high-tech applications like piezoelectric transducers, transparent conducting films and varistors. In this section, the application of ZnO based polymer nanocomposites in sensors and solar cells has been reviewed in brief.

Sensors

Several efforts have been made for the development of different features of sensors by consolidating the organic materials with the inorganic counterparts [101,102]. Subsequently, different inorganic-organic nanocomposites based sensors have been developed [103,104]. These materials consolidate the flexibility and processability of polymers with the properties of the inorganic fillers. Nevertheless, because of the long term instability at raised temperature, it is advantageous to generate sensors which could perform at room temperature.

Nanocrystalline ZnO is a standout amongst the most interesting and broadly utilized inorganic materials for the sensing of LPG, NH_3, H_2 and NO_2 gases [105]. Arshak *et al.* [106] reported the development of a thick-film ZnO/ ZnFe$_2$O$_4$ sensor, which works at room temperature and a drop-coated CPC sensor consisting of 30 w/w% ZnO/ZnFe$_2$O$_4$ nanoparticles. The aforementioned sensors were examined in a completely automated test rig and demonstrated the potential for alcohol vapor sensing.

Solar Cells

The organic photovoltaics (OPVs) have remarkably advanced with respect to their power conversion efficiencies (PCEs) because of the modern improvements in their device architecture, interfacial layers, and active layer materials [107]. In spite of the fact that the operating layer material (for example conjugated polymers) critically affects the power conversion efficiencies, the electron transport layer (ETL), utilized as an interfacial layer between the cathode and operating layer, additionally performs a vital role since it can set up ohmic contact, alter inherent electric field as well as limit the rate of carrier recombination in organic photovoltaics. In the inverted OPV devices, ZnO is utilized as an electron transfer layer because of its greater transparency to visible light, air-stability, adjustable electro-optical properties and higher high electron affinity [108]. Additionally, organic polymers consisting of simple aliphatic amine groups, like polyethylenimine (PEI) and polyethylenimine ethoxylated (PEIE), have also been utilized as electron transfer layers to diminish the operating capacity of ITO as well as lessen the interfacial energy obstruction for the electron transportation from the operating layer to the electrode approaching ohmic contact by generating an interfacial dipole between ETL and operating layer. Despite the fact that ZnO is an appropriate material for electron transfer layer, the use of ZnO nanoparticles can prompt trapping of electrons as well as a greater series resistance due to the presence of defects/traps with oxygen adsorbed on the surface of nanoparticles, thus, indicating that the use of ZnO/polymer composites may be more beneficial.

In a study by Chen *et al.* [109], PEI doped sol-gel prepared ZnO nanocomposites were used as effective electron transport layer in inverted polymer solar cells for assisting electron extraction. The authors reported that the energy bands of PEI/ZnO nanocomposite films could be adjusted by differing the PEI content till 7 wt%. Thus, the conduction band extended from 4.32 eV to 4.0 eV and ZnO structural order in the PEI/ZnO nanocomposite thin films improved for aligning at right angles to the indium tin oxide cathode, promoting the vertical electron transportation. For PEI:ZnO (7:93, w/w) composite utilized as the electron transfer layer, the PCEs of the poly(3-hexylthiophene:phenyl-C61-butryric corrosive methyl ester (1:1, w/w) device enhanced to 4.6% from 3.7% for the analogous device which included pure ZnO as the ETL - a comparative increment of 24%. Figure 6.8 also demonstrates the atomic force micrographs of the ZnO/PEI blend films at different PEI contents.

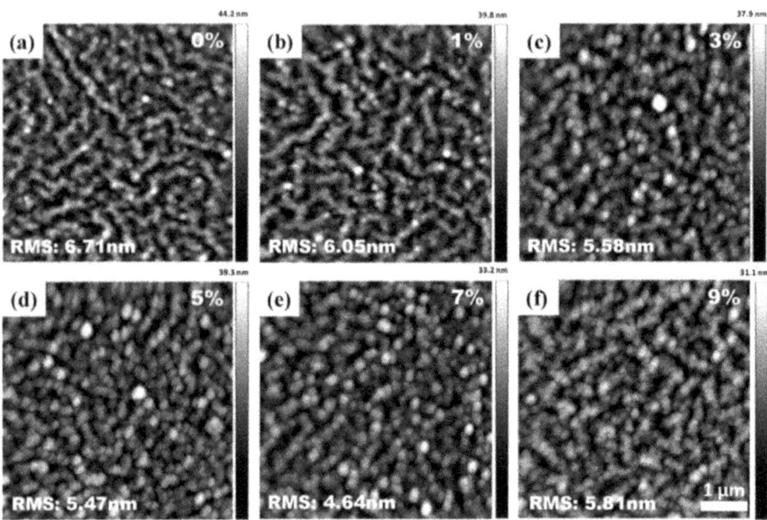

Figure 6.8 Atomic force micrographs of the ZnO/PEI blend films with PEI contents of (a) 0, (b) 1, (c) 3, (d) 5, (e) 7, and (f) 9 wt%. Reproduced from Reference 109 with permission from American Chemical Society.

6.3 Summary

In this chapter, the developments and prospects of different polymer nanocomposites for various electronics applications have been summarized. Specifically, the electronics applications of polymer nanocomposites with nanofillers such as graphene, CNTs, CNFs and ZnO have been reviewed. Currently, the dimensional scaling for the electronic devices has entered the nano-range. Nanotechnology is profoundly embedded in the design of state-of-the-art devices for different electronics as well as opto-electronic applications. The use of polymer-based nanocomposites in these fields is very promising. In this respect, the polymer nanocomposites have either been utilized or have exhibited the potential of utilization in supercapacitors, photovoltaic (PV) cells, photoconductive (PC) cells, fuel cells, electroluminescent devices, chemical sensors, batteries, electro-catalysis and memory devices. Other promising applications include light emitting diodes, printable conductors and field effect transistors.

References

1. Liu, Q., Chen, J., Li, Y., and Shi, G. (2016) High-performance strain sensors with fish-scale-like graphene-sensing layers for full-range detection of human motions. *ACS Nano*, **10**(8), 7901-7906.
2. Yan, C., Wang, J., Kang, W., Cui, M., Wang, X., Foo, C. Y., Chee, K. J., and Lee, P. S. (2014) Highly stretchable piezoresistive graphene-nanocellulose nanopaper for strain sensors. *Advanced Materials*, **26**(13), 2022-2027.
3. Schwartz, G., Tee, B. C., Mei, J., Appleton, A. L., Kim, D. H., Wang, H., and Bao, Z. (2013) Flexible polymer transistors with high pressure sensitivity for application in electronic skin and health monitoring. *Nature Communications*, **4**, 1859.
4. Crone, B., Dodabalapur, A., Gelperin, A., Torsi, L., Katz, H. E., Lovinger, A. J., and Bao, Z. (2001) Electronic sensing of vapors with organic transistors. *Applied Physics Letters*, **78**(15), 2229-2231.
5. Crone, B. K., Dodabalapur, A., Sarpeshkar, R., Filas, R. W., Lin, Y. Y., Bao, Z., O'Neill, J. H., Li, W., and Katz, H. E. (2001) Design and fabrication of organic complementary circuits. *Journal of Applied Physics*, **89**(9), 5125-5132.
6. Boland, C. S., Khan, U., Ryan, G., Barwich, S., Charifou, R., Harvey, A., Backes, C., Li, Z., Ferreira, M. S., Mobius, M. E., and Young, R. J. (2016) Sensitive electromechanical sensors using viscoelastic graphene-polymer nanocomposites. *Science*, **354**(6317), 1257-1260.
7. Hu, C. H., Liu, C. H., Chen, L. Z., Peng, Y. C., and Fan, S. S. (2008) Resistance-pressure sensitivity and a mechanism study of multiwall carbon nanotube networks/poly (dimethylsiloxane) composites. *Applied Physics Letters*, **93**(3), 033108.
8. Zhang, R., Baxendale, M., and Peijs, T. (2007) Universal resistivity-strain dependence of carbon nanotube/polymer composites. *Physical Review B*, **76**(19), 195433.
9. Toprakci, H. A., Kalanadhabhatla, S. K., Spontak, R. J., and Ghosh, T. K. (2013) Polymer nanocomposites containing carbon nanofibers as soft printable sensors exhibiting strain-reversible piezoresistivity. *Advanced Functional Materials*, **23**(44), 5536-5542.
10. Borghetti, M., Serpelloni, M., Sardini, E., and Pandini, S. (2016) Mechanical behavior of strain sensors based on PEDOT:PSS and silver

nanoparticles inks deposited on polymer substrate by inkjet printing. *Sensors and Actuators A: Physical*, **243**, 71-80.

11. Friend, R. H., Gymer, R. W., Holmes, A. B., and Burroughes, J. H. (1999) Electroluminescence in conjugated polymers. *Nature*, **397**(6715), 121-128.

12. Anglos, D., Stassinopoulos, A., Das, R. N., Zacharakis, G., Psyllaki, M., Jakubiak, R., Vaia, R. A., Giannelis, E. P., and Anastasiadis, S. H. (2004) Random laser action in organic–inorganic nanocomposites. *Journal of the Optical Society of America B*, **21**(1), 208-213.

13. Blanchet, G. B., Fincher, C. R., and Gao, F. (2003) Polyaniline nanotube composites: a high-resolution printable conductor. *Applied Physics Letters*, **82**(8), 1290-1292.

14. Huang, D., Liao, F., Molesa, S., Redinger, D., and Subramanian, V. (2003) Plastic-compatible low resistance printable gold nanoparticle conductors for flexible electronics. *Journal of the Electrochemical Society*, **150**(7), G412-G417.

15. Lee, H. H., Chou, K. S., and Huang, K. C. (2005) Inkjet printing of nanosized silver colloids. *Nanotechnology*, **16**(10), 2436-2441.

16. Wu, Y., Li, Y., Liu, P., Gardner, S., and Ong, B. S. (2006) Studies of gold nanoparticles as precursors to printed conductive features for thin-film transistors. *Chemistry of Materials*, **18**(19), 4627-4632.

17. Forrest, S. R. (2004) The path to ubiquitous and low-cost organic electronic appliances on plastic. *Nature*, **428**(6986), 911-918.

18. Amarasekera, J. (2005) Conductive plastics for electrical and electronic applications. *Reinforced Plastics*, **49**(8), 38-41.

19. Byrne, M. T., and Gun'ko, Y. K. (2010) Recent advances in research on carbon nanotube-polymer composites. *Advanced Materials*, **22**(15), 1672-1688.

20. Xie, X. L., Mai, Y. W., and Zhou, X. P. (2005) Dispersion and alignment of carbon nanotubes in polymer matrix: A review. *Materials Science and Engineering R: Reports*, **49**(4), 89-112.

21. Sadasivuni, K. K., Ponnamma, D., Thomas, S., and Grohens, Y. (2014) Evolution from graphite to graphene elastomer composites. *Progress in Polymer Science*, **39**(4), 749-780.

22. Lu, C. C., Lin, Y. C., Yeh, C. H., Huang, J. C., and Chiu, P. W. (2012) High mobility flexible graphene field-effect transistors with self-healing gate dielectrics. *ACS Nano*, **6**(5), 4469-4474.

23. Liu, Z., Liu, Q., Huang, Y., Ma, Y., Yin, S., Zhang, X., Sun, W., and Chen, Y. (2008) Organic photovoltaic devices based on a novel acceptor material: graphene. *Advanced Materials*, **20**(20), 3924-3930.

24. Barber, P., Balasubramanian, S., Anguchamy, Y., Gong, S., Wibowo, A., Gao, H., Ploehn, H. J., and Zur Loye, H. C. (2009) Polymer composite and nanocomposite dielectric materials for pulse power energy storage. *Materials*, **2**(4), 1697-1733.

25. Bolotin, K. I., Sikes, K. J., Jiang, Z., Klima, M., Fudenberg, G., Hone, J.,

Kim, P., and Stormer, H. L. (2008) Ultrahigh electron mobility in suspended graphene. *Solid State Communications*, **146**(9), 351-355.

26. Wu, Q., Xu, Y., Yao, Z., Liu, A., and Shi, G. (2010) Supercapacitors based on flexible graphene/polyaniline nanofiber composite films. *ACS Nano*, **4**(4), 1963-1970.

27. Yan, J., Wei, T., Shao, B., Fan, Z., Qian, W., Zhang, M., and Wei, F. (2010) Preparation of a graphene nanosheet/polyaniline composite with high specific capacitance. *Carbon*, **48**(2), 487-493.

28. Xu, C., Wang, X., and Zhu, J. (2008) Graphene-metal particle nanocomposites. *The Journal of Physical Chemistry C*, **112**(50), 19841-19845.

29. Sridhar, P., Perumal, R., Rajalakshmi, N., Raja, M., and Dhathathreyan, K. S. (2001) Humidification studies on polymer electrolyte membrane fuel cell. *Journal of Power Sources*, **101**(1), 72-78.

30. Kim, J. R., Cheng, S., Oh, S. E., and Logan, B. E. (2007) Power generation using different cation, anion, and ultrafiltration membranes in microbial fuel cells. *Environmental Science & Technology*, **41**(3), 1004-1009.

31. Wang, J., Wang, Y., He, D., Wu, H., Wang, H., Zhou, P., and Fu, M. (2012) Influence of polymer/fullerene-graphene structure on organic polymer solar devices. *Integrated Ferroelectrics*, **137**(1), 1-9.

32. Hong, W., Xu, Y., Lu, G., Li, C., and Shi, G. (2008) Transparent graphene/PEDOT–PSS composite films as counter electrodes of dye-sensitized solar cells. *Electrochemistry Communications*, **10**(10), 1555-1558.

33. Higashihara, T., Matsumoto, K., and Ueda, M. (2009) Sulfonated aromatic hydrocarbon polymers as proton exchange membranes for fuel cells. *Polymer*, **50**(23), 5341-5357.

34. Dreyer, D. R., Park, S., Bielawski, C. W., and Ruoff, R. S. (2010) The chemistry of graphene oxide. *Chemical Society Reviews*, **39**(1), 228-240.

35. Liu, S., Zeng, T. H., Hofmann, M., Burcombe, E., Wei, J., Jiang, R., Kong, J., and Chen, Y. (2011) Antibacterial activity of graphite, graphite oxide, graphene oxide, and reduced graphene oxide: Membrane and oxidative stress. *ACS Nano*, **5**(9), 6971-6980.

36. Mayer, A. C., Scully, S. R., Hardin, B. E., Rowell, M. W., and McGehee, M. D. (2007) Polymer-based solar cells. *Materials Today*, **10**(11), 28-33.

37. Liu, J., Qiao, Y., Guo, C. X., Lim, S., Song, H., and Li, C. M. (2012) Graphene/carbon cloth anode for high-performance mediatorless microbial fuel cells. *Bioresource Technology*, **114**, 275-280.

38. Yong, Y. C., Dong, X. C., Chan-Park, M. B., Song, H., and Chen, P. (2012) Macroporous and monolithic anode based on polyaniline hybridized three-dimensional graphene for high-performance microbial

fuel cells. *ACS Nano*, **6**(3), 2394-2400.

39. Hsu, C. L., Lin, C. T., Huang, J. H., Chu, C. W., Wei, K. H., and Li, L. J. (2012) Layer-by-layer graphene/TCNQ stacked films as conducting anodes for organic solar cells. *ACS Nano*, **6**(6), 5031-5039.

40. Tang, C. W. (1986) Two-layer organic photovoltaic cell. *Applied Physics Letters*, **48**(2), 183-185.

41. Alam, M. M., and Jenekhe, S. A. (2004) Efficient solar cells from layered nanostructures of donor and acceptor conjugated polymers. *Chemistry of Materials*, **16**(23), 4647-4656.

42. Chang, H., Liu, Y., Zhang, H., and Li, J. (2011) Pyrenebutyrate-functionalized graphene/poly (3-octyl-thiophene) nanocomposites based photoelectrochemical cell. *Journal of Electroanalytical Chemistry*, **656**(1), 269-273.

43. Muhlbacher, D., Scharber, M., Morana, M., Zhu, Z., Waller, D., Gaudiana, R., and Brabec, C. (2006) High photovoltaic performance of a low-bandgap polymer. *Advanced Materials*, **18**(21), 2884-2889.

44. Eswaraiah, V., Balasubramaniam, K., and Ramaprabhu, S. (2011) Functionalized graphene reinforced thermoplastic nanocomposites as strain sensors in structural health monitoring. *Journal of Materials Chemistry*, **21**(34), 12626-12628.

45. Sadek, A. Z., Wlodarski, W., Kalantar-Zadeh, K., Baker, C., and Kaner, R. B. (2007) Doped and dedoped polyaniline nanofiber based conductometric hydrogen gas sensors. *Sensors and Actuators A: Physical*, **139**(1), 53-57.

46. Al-Mashat, L., Shin, K., Kalantar-zadeh, K., Plessis, J. D., Han, S. H., Kojima, R. W., Kaner, R. B., Li, D., Gou, X., Ippolito, S. J., and Wlodarski, W. (2010) Graphene/polyaniline nanocomposite for hydrogen sensing. *The Journal of Physical Chemistry C*, **114**(39), 16168-16173.

47. Eswaraiah, V., Balasubramaniam, K., and Ramaprabhu, S. (2012) One-pot synthesis of conducting graphene–polymer composites and their strain sensing application. *Nanoscale*, **4**(4), 1258-1262.

48. Dresselhaus, M. S., Dresselhaus, G., and Saito, R. (1995) Physics of carbon nanotubes. *Carbon*, **33**(7), 883-891.

49. Baughman, R. H., Zakhidov, A. A., and De Heer, W. A. (2002) Carbon nanotubes - the route toward applications. *Science*, **297**(5582), 787-792.

50. Lee, S. H., Lee, D. H., Lee, W. J., and Kim, S. O. (2011) Tailored assembly of carbon nanotubes and graphene. *Advanced Functional Materials*, **21**(8), 1338-1354.

51. Coleman, J. N., Khan, U., Blau, W. J., and Gun'ko, Y. K. (2006) Small but strong: a review of the mechanical properties of carbon nanotube–polymer composites. *Carbon*, **44**(9), 1624-1652.

52. Frackowiak, E., Khomenko, V., Jurewicz, K., Lota, K., and Beguin, F. (2006) Supercapacitors based on conducting polymers/nanotubes composites. *Journal of Power Sources*, **153**(2), 413-418.

53. Li, J., Cheng, X., Sun, J., Brand, C., Shashurin, A., Reeves, M., and Keidar, M. (2014) Paper based ultra-capacitors with carbon nanotubes-graphene composites. *Journal of Applied Physics*, **115**(16), 164301.

54. Al-Saleh, M. H., and Sundararaj, U. (2009) Electromagnetic interference shielding mechanisms of CNT/polymer composites. *Carbon*, **47**(7), 1738-1746.

55. Lee, J. I., Yang, S. B., and Jung, H. T. (2009) Carbon nanotubes-polypropylene nanocomposites for electrostatic discharge applications. *Macromolecules*, **42**(21), 8328-8334.

56. Yu, M., He, Q., Yu, D., Zhang, X., Ji, A., Zhang, H., Guo, C., and Dai, Z. (2012) Efficient active actuation to imitate locomotion of gecko's toes using an ionic polymer-metal composite actuator enhanced by carbon nanotubes. *Applied Physics Letters*, **101**(16), 163701.

57. Yadav, S. K., Kim, I. J., Kim, H. J., Kim, J., Hong, S. M., and Koo, C. M. (2013) PDMS/MWCNT nanocomposite actuators using silicone functionalized multiwalled carbon nanotubes via nitrene chemistry. *Journal of Materials Chemistry C*, **1**(35), 5463-5470.

58. Kazaoui, S., Minami, N., Nalini, B., Kim, Y., and Hara, K. (2005) Near-infrared photoconductive and photovoltaic devices using single-wall carbon nanotubes in conductive polymer films. *Journal of Applied Physics*, **98**(8), 084314.

59. Hwang, J., Jang, J., Hong, K., Kim, K. N., Han, J. H., Shin, K., and Park, C. E. (2011) Poly (3-hexylthiophene) wrapped carbon nanotube/poly (dimethylsiloxane) composites for use in finger-sensing piezoresistive pressure sensors. *Carbon*, **49**(1), 106-110.

60. Potschke, P., Kobashi, K., Villmow, T., Andres, T., Paiva, M. C., and Covas, J. A. (2011) Liquid sensing properties of melt processed polypropylene/poly (ε-caprolactone) blends containing multiwalled carbon nanotubes. *Composites Science and Technology*, **71**(12), 1451-1460.

61. Kim, B., Park, Y. D., Min, K., Lee, J. H., Hwang, S. S., Hong, S. M., Kim, B. H., Kim, S. O., and Koo, C. M. (2011) Electric actuation of nanostructured thermoplastic elastomer gels with ultralarge electrostriction coefficients. *Advanced Functional Materials*, **21**(17), 3242-3249.

62. Pelrine, R., Kornbluh, R., Pei, Q., and Joseph, J. (2000) High-speed electrically actuated elastomers with strain greater than 100%. *Science*, **287**(5454), 836-839.

63. Huang, S., Woodson, M., Smalley, R., and Liu, J. (2004) Growth mechanism of oriented long single walled carbon nanotubes using "fast-heating" chemical vapor deposition process. *Nano Letters*, **4**(6), 1025-1028.

64. Jeong, Y. G., and Jeon, G. W. (2013) Microstructure and performance of multiwalled carbon nanotube/m-aramid composite films as

electric heating elements. *ACS Applied Materials & Interfaces*, **5**(14), 6527-6534.

65. Mastragostino, M., Arbizzani, C., and Soavi, F. (2001) Polymer-based supercapacitors. *Journal of Power Sources*, **97**, 812-815.

66. Arbizzani, C., Mastragostino, M., and Soavi, F. (2001) New trends in electrochemical supercapacitors. *Journal of Power Sources*, **100**(1), 164-170.

67. Markham, D. (1999) Shielding: quantifying the shielding requirements for portable electronic design and providing new solutions by using a combination of materials and design. *Materials & Design*, **21**(1), 45-50.

68. Chung, D. D. L. (2001) Electromagnetic interference shielding effectiveness of carbon materials. *Carbon*, **39**(2), 279-285.

69. Huang, J. C. (1995) EMI shielding plastics: a review. *Advances in Polymer Technology*, **14**(2), 137-150.

70. Jiang, G., Gilbert, M., Hitt, D. J., Wilcox, G. D., and Balasubramanian, K. (2002). Preparation of nickel coated mica as a conductive filler. *Composites, Part A: Applied Science and Manufacturing*, **33**(5), 745-751.

71. Chung, D. D. L. (2004) Electrical applications of carbon materials. *Journal of Materials Science*, **39**(8), 2645-2661.

72. Li, N., Huang, Y., Du, F., He, X., Lin, X., Gao, H., Ma, Y., Li, F., Chen, Y., and Eklund, P. C. (2006) Electromagnetic interference (EMI) shielding of single-walled carbon nanotube epoxy composites. *Nano Letters*, **6**(6), 1141-1145.

73. Jou, W. S., Cheng, H. Z., and Hsu, C. F. (2007) The electromagnetic shielding effectiveness of carbon nanotubes polymer composites. *Journal of Alloys and Compounds*, **434**, 641-645.

74. Yang, Y., Gupta, M. C., Dudley, K. L., and Lawrence, R. W. (2007) Electromagnetic interference shielding characteristics of carbon nanofiber-polymer composites. *Journal of Nanoscience and Nanotechnology*, **7**(2), 549-554.

75. Yang, S., Lozano, K., Lomeli, A., Foltz, H. D., and Jones, R. (2005) Electromagnetic interference shielding effectiveness of carbon nanofiber/LCP composites. *Composites, Part A: Applied Science and Manufacturing*, **36**(5), 691-697.

76. Schulz, R. B., Plantz, V. C., and Brush, D. R. (1988) Shielding theory and practice. *IEEE Transactions on Electromagnetic Compatibility*, **30**(3), 187-201.

77. Yang, Y., Gupta, M. C., Dudley, K. L., and Lawrence, R. W. (2005) Conductive carbon nanofiber-polymer foam structures. *Advanced Materials*, **17**(16), 1999-2003.

78. Gojny, F. H., Wichmann, M. H., Fiedler, B., Kinloch, I. A., Bauhofer, W., Windle, A. H., and Schulte, K. (2006) Evaluation and identification of electrical and thermal conduction mechanisms in carbon nanotube/

epoxy composites. *Polymer*, **47**(6), 2036-2045.

79. Das, N. C., Chaki, T. K., and Khastgir, D. (2002) Effect of processing parameters, applied pressure and temperature on the electrical resistivity of rubber-based conductive composites. *Carbon*, **40**(6), 807-816.

80. Dang, Z. M., Jiang, M. J., Xie, D., Yao, S. H., Zhang, L. Q., and Bai, J. (2008) Supersensitive linear piezoresistive property in carbon nanotubes/silicone rubber nanocomposites. *Journal of Applied Physics*, **104**(2), 024114.

81. Yu, M. F., Lourie, O., Dyer, M. J., Moloni, K., Kelly, T. F., and Ruoff, R. S. (2000) Strength and breaking mechanism of multiwalled carbon nanotubes under tensile load. *Science*, **287**(5453), 637-640.

82. Roldughin, V. I., and Vysotskii, V. V. (2000) Percolation properties of metal-filled polymer films, structure and mechanisms of conductivity. *Progress in Organic Coatings*, **39**(2), 81-100.

83. Hussain, M., Choa, Y.-H., and Niihara, K. (2001) Fabrication process and electrical behavior of novel pressure-sensitive composites. *Composites, Part A: Applied Science and Manufacturing*, **32**(12), 1689-1696.

84. Athanassiou, E. K., Krumeich, F., Grass, R. N., and Stark, W. J. (2008) Advanced piezoresistance of extended metal-insulator core-shell nanoparticle assemblies. *Physical Review Letters*, **101**(16), 166804.

85. Kang, I., Schulz, M. J., Kim, J. H., Shanov, V., and Shi, D. (2006) A carbon nanotube strain sensor for structural health monitoring. *Smart Materials and Structures*, **15**(3), 737.

86. Pham, G. T., Colombo, A., Park, Y. B., Zhang, C., and Wang, B. (2006) Nanotailored Thermoplastic/Carbon Nanotube Composite Strain Sensor. In: *Proceedings of the Multifunctional Nanocomposites International Conference*, USA, pp. 277-283.

87. Wang, P., Geng, S., and Ding, T. (2010) Effects of carboxyl radical on electrical resistance of multi-walled carbon nanotube filled silicone rubber composite under pressure. *Composites Science and Technology*, **70**(10), 1571-1573.

88. Collins, P. G., Bradley, K., Ishigami, M., and Zettl, A. (2000) Extreme oxygen sensitivity of electronic properties of carbon nanotubes. *Science*, **287**(5459), 1801-1804.

89. McDonald, S. A., Konstantatos, G., Zhang, S., Cyr, P. W., Klem, E. J., Levina, L., and Sargent, E. H. (2005) Solution-processed PbS quantum dot infrared photodetectors and photovoltaics. *Nature Materials*, **4**(2), 138-142.

90. Avouris, P., Chen, J., Freitag, M., Perebeinos, V., and Tsang, J. C. (2006) Carbon nanotube optoelectronics. *Physica Status Solidi (b)*, **243**(13), 3197-3203.

91. Pradhan, B., Setyowati, K., Liu, H., Waldeck, D. H., and Chen, J. (2008) Carbon nanotube-polymer nanocomposite infrared sensor. *Nano*

Letters, **8**(4), 1142-1146.

92. Halls, J. J. M., Walsh, C. A., Greenham, N. C., Marseglia, E. A., Friend, R. H., Moratti, S. C., and Holmes, A. B. (1995) Efficient photodiodes from interpenetrating polymer networks. *Nature*, **376**(6540), 498-500.

93. Bhattacharyya, S., Kymakis, E., and Amaratunga, G. A. J. (2004) Photovoltaic properties of dye functionalized single-wall carbon nanotube/conjugated polymer devices. *Chemistry of Materials*, **16**(23), 4819-4823.

94. Arranz-Andres, J., and Blau, W. J. (2008) Enhanced device performance using different carbon nanotube types in polymer photovoltaic devices. *Carbon*, **46**(15), 2067-2075.

95. Miyagawa, H., Rich, M. J., and Drzal, L. T. (2006) Thermo-physical properties of epoxy nanocomposites reinforced by carbon nanotubes and vapor grown carbon fibers. *Thermochimica Acta*, **442**(1), 67-73.

96. Luo, X., and Chung, D. D. L. (1999) Electromagnetic interference shielding using continuous carbon-fiber carbon-matrix and polymer-matrix composites. *Composites, Part B: Engineering*, **30**(3), 227-231.

97. Wu, J., and Chung, D. D. L. (2002) Increasing the electromagnetic interference shielding effectiveness of carbon fiber polymer–matrix composite by using activated carbon fibers. *Carbon*, **40**(3), 445-447.

98. Katsumata, M., Yamanashi, H., Ushijima, H., and Endo, M. (1993) EMI Application of Highly Conductive Plastic Composite using Vapor-Grown Carbon Fibers (VGCF). *1993 North American Conference on Smart Structures and Materials, International Society for Optics and Photonics*, pp. 140-148.

99. Wu, S., Zhang, J., Ladani, R. B., Ravindran, A. R., Mouritz, A. P., Kinloch, A. J., and Wang, C. H. (2017) Novel electrically conductive porous PDMS/carbon nanofiber composites for deformable strain sensors and conductors. *ACS Applied Materials & Interfaces*, **9**(16), 14207-14215.

100. He, Y. (2005) A novel emulsion route to sub-micrometer polyaniline/nano-ZnO composite fibers. *Applied Surface Science*, **249**(1), 1-6.

101. Tiwari, A., Mishra, A. P., Dhakate, S. R., Khan, R., and Shukla, S. K. (2007) Synthesis of electrically active biopolymer–SiO_2 nanocomposite aerogel. *Materials Letters*, **61**(23), 4587-4590.

102. Tiwari, A., and Shukla, S. K. (2009) Chitosan-g-polyaniline: a creatine amidinohydrolase immobilization matrix for creatine biosensor. *Express Polymer Letters*, **3**(9), 553-559.

103. Chang, C. P., and Yuan, C. L. (2009) The fabrication of a MWNTs–polymer composite chemoresistive sensor array to discriminate between chemical toxic agents. *Journal of Materials Science*, **44**(20),

5485-5493.
104. Koul, S., Chandra, R., and Dhawan, S. K. (2001) Conducting polyaniline composite: a reusable sensor material for aqueous ammonia. *Sensors and Actuators B: Chemical*, **75**(3), 151-159.
105. Bahnemann, D. W., Kormann, C., and Hoffmann, M. R. (1987) Preparation and characterization of quantum size zinc oxide: a detailed spectroscopic study. *Journal of Physical Chemistry*, **91**(14), 3789-3798.
106. Arshak, K., Moore, E., Cunniffe, C., Nicholson, M., and Arshak, A. (2007) Preparation and characterisation of $ZnFe_2O_4$/ZnO polymer nanocomposite sensors for the detection of alcohol vapours. *Superlattices and Microstructures*, **42**(1), 479-488.
107. Steim, R., Kogler, F. R., and Brabec, C. J. (2010) Interface materials for organic solar cells. *Journal of Materials Chemistry*, **20**(13), 2499-2512.
108. Park, H. Y., Lim, D., Kim, K. D., and Jang, S. Y. (2013) Performance optimization of low-temperature-annealed solution-processable ZnO buffer layers for inverted polymer solar cells. *Journal of Materials Chemistry A*, **1**(21), 6327-6334.
109. Chen, H. C., Lin, S. W., Jiang, J. M., Su, Y. W., and Wei, K. H. (2015) Solution-processed zinc oxide/polyethylenimine nanocomposites as tunable electron transport layers for highly efficient bulk heterojunction polymer solar cells. *ACS Applied Materials & Interfaces*, **7**(11), 6273-6281.

Polymer Nanocomposites for Sensing Applications

7.1 Introduction

Sensors are devices which are capable of providing information about physical, chemical or biological environments. For instance, there is a need to sense and control toxic gases and vapors released from industries to the atmosphere. Similarly, industrial effluents released in natural water bodies are required to be monitored. These needs have paved the way for the development of various chemical, biological and physical sensors [1]. The physical sensors measure physical quantities and convert these into a signal. A functional definition for a chemical sensor is "a small device which, as the result of a chemical interaction or process between the analyte gas and sensor device, transforms chemical or biochemical information (quantitative or qualitative) into an analytically useful signal" [2]. A chemical sensor provides information about its environment and consists of a chemically selective layer and a physical transducer [3-7]. On the other hand, a biosensor is different from chemical sensor as it has a biological entity like tissue, bacteria, antibody, enzyme, etc., acting as a recognition agent. These biologically sensitive entities can be generated by employing the biological engineering strategies. The interaction of the analyte with the biological element is detected by the biosensor and transformed into a more quantifiable and easily measurable signal using a transducer which works in electrochemical, piezoelectric or optical mode [7-11], thus, representing a functional system.

In the recent years, different types of materials have been utilized to generate advanced sensors, however, polymers based sensors have gained particular research interest as the physical and chemical properties of polymers can be effectively tailored to achieve the superior sensor performance [11-19].

In this respect, this review aims at summarizing the recent developments in the development of the polymer composite systems for use in various sensing applications.

Liyamol Jacob and Vikas Mittal, Khalifa University of Science and Technology, Abu Dhabi, UAE

7.2 Conducting Polymer Nanocomposites for Sensing

Conducting polymer nanocomposites are a class of hybrid nano-materials, generated by incorporating various conducting nano-fillers into different polymers. Thus, the composites combine the intrinsic properties of both the polymer matrix and reinforcement [20]. Nanowires, nanotubes as well as nanoparticles of various metals like gold and silver have been used as conducting nanosized filler materials in various polymers [21-33]. Among these, carbon nanotubes (CNTs) have received vast applications in chemical sensing, biosensing, nano-electronic devices, etc. [34,35]. Superior stability, mechanical performance, surface area and electrical conductivity make the CNTs a good candidate for generating polymer nanocomposites for sensing applications [36-41]. Highly effective enzyme loading, large surface area per unit mass and insignificant diffusion limitations can be achieved by using metal nanoparticles [42]. Electrochemical biosensors and chemical sensors are, thus, widely studied with metal nanoparticles embedded polymer nanocomposites [43-45].

Conducting polymers also exhibit unique electrical, optical, mechanical, magnetic and electronic properties which drive their use as conducting nanocomposites in a wide range of applications like single electron transistors, electrochemical devices, biosensors and nanotips of field emission displays because of the nanoscale π-conjugated organic molecules [46-56].

The easy installation and usage of mobile sensors for various gases like hydrocarbons, nitrates, carbon monoxide, etc., are of great importance. Low power consuming non-intrusive sensors, which are cost effective and exhibit rapid response towards the molecules, need to be designed. The device should also be able to detect very minute level of contamination. For instance, CO gas sensors based on ultrathin conducting films with metal oxides like SnO_2 and TiO_2 were reported [57]. Layer by layer self-assembly technique was used to fabricate the polymer-metal oxide nanocomposites.

There is also a rising demand for generating sensors which are wearable, flexible and stretchable, and can be attached to human body directly or mounted onto clothes owing to their facile interaction with the human body. For detecting the motion in humans, stretchable and sensitive strain sensors can be introduced. For instance, sandwiched polymer nanocomposite made of poly(dimethylsiloxane) (PDMS) and silver nanowires were employed to generate ultra-sensitive strain sensors (Figure 7.1). The nanocomposite based sensors exhibited a high stretchability of around 70% and high piezoelectric resistivity with tunable gauge factors in the range of 2 to 14. The authors successfully demonstrated the fabrication of a glove integrated with five strain sensors for the detection of the finger movement.

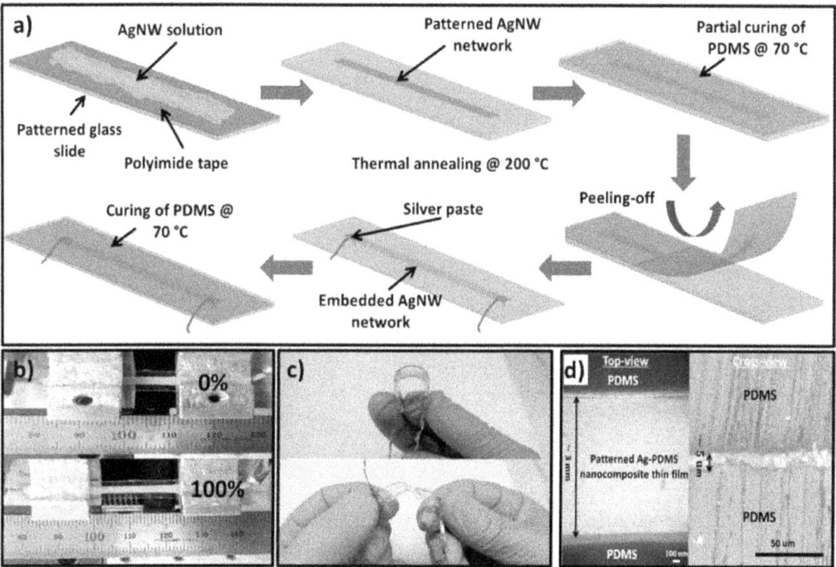

Figure 7.1 Fabrication scheme and performance of the sandwich-structured PDMS/silver nanowires/PDMS nanocomposite strain sensor. Reproduced from Reference 58 with permission from American Chemical Society.

7.3 Development of Different Sensors

7.3.1 Gas Sensors

Human health is under ever increasing threat because of the rapidly enhancing environmental pollution and release of harmful gases in the environment. Numerous efforts have been made to develop sensors which can detect the gas molecules posing a hazard for humans and environment. Using such sensitive, selective and efficient gas sensors, explosive materials (at approx. 5 ppm), chemical vapors (at approx. 1 ppb) and gases (at ppt level) can be detected. A schematic of the mechanism and detection types has been presented in Figure

7.2 [59]. For pollution monitoring and toxic gases or vapors detection, miniaturized gas sensors utilizing polymer composites have received widespread recognition in the recent past.

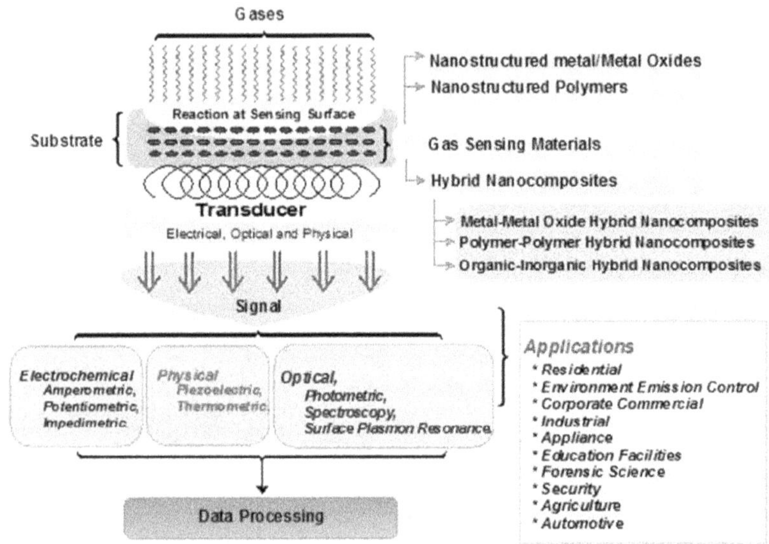

Figure 7.2 Schematic of gas sensor fabrication techniques. Reproduced from Reference 59 with permission from American Chemical Society.

Abraham *et al.* [60] reported compact wireless gas sensors fabricated using surface modified CNTs/poly(methyl methacrylate) (PMMA) composite film. With the exposure to different gases, there was a change in the resistance of the composite film which was used as the principle to detect the presence of gases. The composite film showed fast response and change in resistance of the order of 10^2-10^3 for different organic vapors. In another study, Tai *et al.* [61] reported *in-situ* chemical oxidation polymerization approach to generate polyaniline-titanium dioxide (PANI/TiO$_2$) nanocomposites. The composite film was mounted onto a silicon substrate (using self-assembly approach) covered with interdigital electrodes to generate a gas sensor. The gas detection capability of the composite was tested against CO and NH$_3$. Compared to CO, the stability, reproducibility and response was superior for NH$_3$ gas. Humidity had less effects on the resistance of the nanocomposite film. The authors also compared the performance of the nanocomposite film with pure PANI and observed differences in both gas-sensing and surface morphology. The authors also studied the effect of processing temperature on the gas response of the PANI/TiO$_2$ thin film gas sensor [62]. The composites prepared at 10 °C exhibited superior reproducibility, sensitivity, selectivity and stability as compared to the polymer composites generated using other temperatures.

Suri *et al.* [63] employed gelation and polymerization simultaneously for generating nanocomposites of iron oxide and polypyrrole (PPy). At higher polypyrrole concentration, iron oxide was present as a single cubic phase, while at lower polypyrrole concentration, it was present in a mixed phase. Both gas and humidity sensing properties of the composites were studied. The increasing concentration of polypyrrole enhanced the humidity sensitivity of the nanocomposites. The pressure and sensitivity were related linearly for studied gases like CO$_2$, N$_2$ and CH$_4$. In another study, Mishra *et al.* [64] reported surface plasmon resonance phenomenon based fiber optic gas sensors utilizing PMMA and reduced graphene oxide (rGO). The unclad portion of the optical fiber was coated with copper film which was further coated with PMMA and PMMA/rGO. Nitrogen, hydrogen, chlorine, hydrogen and ammonia gases were tested for the applicability of the polymer composite gas sensors. Wavelength interrogation technique was used for detecting the gas sensitivity of the composites for various gases at different concentrations. With the increase in the concentration of the gases in the chamber, the resonance wavelength exhibited a red shift. The authors observed that the probes exhibited higher sensitivity towards ammonia gas. Also, the probe with PMMA/rGO nanocomposite over-layer coating exhibited the best performance. Chen *et al.* [65] also used poly(ethylene-block-ethylene oxide) (PE-b-PEO)-grafted carbon black to generate polyethylene composites with high stability and dispersity. The composites were studied for response against different solvent vapors. In solvents like cyclohexane and carbon tetrachloride at 40 °C, the resistance of the composite changed swiftly by 10^4-10^6 times. After returning the composite to dry air, the resistance returned to its initial state. At the same temperature, for solvents like water and alcohol, the resistance changed only marginally.

Shang *et al.* [66] used micro-emulsion technique to generate gas sensors based on PMMA/carbon nanotubes composites. Compared to the solution mixed samples of PMMA and CNTs, micro-emulsion technique was successful in developing more uniform composite films with enhanced electrical properties. Upon exposure to various gases, the resistance of the composite film was observed to change, which was reversible in nature. In addition, the authors observed higher response and lower percolation threshold for the composite prepared using micro-emulsion technique.

Tandon *et al.* [67] reported nanocomposites of PPy-Fe_3O_4 using emulsion polymerization. The composites had high conductivity, however, increasing the PPy content resulted in a decrease in conductivity. The composites were observed to be sensitive towards humidity and commonly used gases such as N_2, O_2 and CO_2. Yang *et al.* [68] reported deposition of poly(3,4-ethylenedioxythiophene) (PEDOT) on reduced graphene oxide films. An excellent performance was exhibited by the composite in NO_2 sensing due to the synergistic effect of porous structure and large surface area. Pure rGO was also compared against the composite and the sensitivity of the polymer composite was observed to be 2 orders of magnitude higher. The composite also exhibited higher response and recovery performance. Figure 7.3 shows the influence of porous structure on the gas sensitivity.

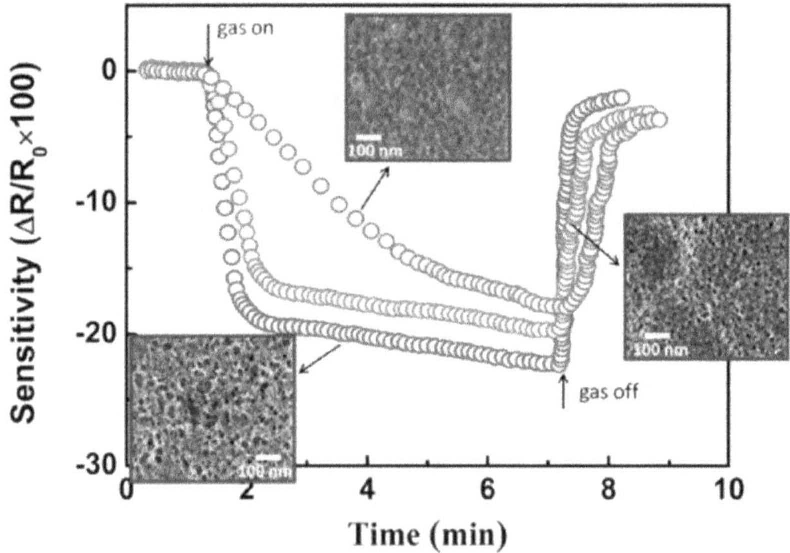

Figure 7.3 Influence of the porous structure on the NO_2 sensitivity (5 ppm). Reproduced from Reference 68 with permission from American Chemical Society.

In another study, Sadek *et al.* [69] reported layered surface acoustic wave (SAW) sensor fabricated using polyaniline/In_2O_3 nanofiber composites. In the presence of finely divided In_2O_3, aniline was subjected to chemical oxidative polymerization to generate polyaniline nanofiber/In_2O_3 nanocomposites. ZnO/64° YX $LiNbO_3$ SAW was used as a transducer to deposit the films generated using this technique. At room temperature, the composites showed good reproducibility, swift recovery and fast response time for different gases like H_2, NO_2 and CO.

7.3.2 Biosensors

Environmental stability, biocompatibility, fast electrode kinetics, monitorable chemical and electrochemical properties, reversibility in doping and dedoping, a large basal plane and edge plane ratio, etc., are some of the benefits of using polymer composites for biosensing [70-75].

Remarkable optical, electronic and catalytic properties of gold nanoparticles (AuNPs) have driven their use in various research areas. AuNPs@molecularly imprinted polymers (MIP) sensors have also been generated by integrating gold nanoparticles with different MIPs [76]. Due to the exceptional chemical properties of the MIPs, it is possible to detect a particular analyte with high sensitivity and electability from a complex mixture of molecules. The recognition event can be changed into a measurable physical event like optical, electrical and piezoelectric, etc. By the presence of gold nanoparticles, the physical properties can be enhanced. Figure 7.4 presents the schematic of the hybrid synthesis. In another study, emulsion polymerization was used to synthesize poly(p-phenylenediamine) (PpPDA)-Fe_3O_4 composite [77]. Hemoglobin (Hb) was immobilized onto the composite and its biological and electrochemical changes were observed. Owing to direct electron transfer between the protein and underlying electrode, well defined redox peaks of Hb were observed. The detection limit exhibited by the biosensor was 0.21 µM (S/N=3). It also presented good sensitivity towards H_2O_2 as well as high reproducibility. Concentration of H_2O_2 was proportional to catalytic reduction current within the range of 0.5-400.0 µM. The high affinity of Hb towards the polymer composite could be confirmed by a Michaelis-Menten constant of 0.088 mM.

Emre *et al.* [78] studied poly(4-(2,3-dihydrothieno[3,4-*b*][1,4]dioxin-5-yl)-7-(2,3-dihydrothieno[3,4-b][1,4]dioxin-7-yl)-2-benzyl-1*H*-benzo[d]imidazole) (poly(BIPE)), which is a conducting polymer and a PMMA-clay nanocomposite with 2-(methacryloyloxy) ethyltri-

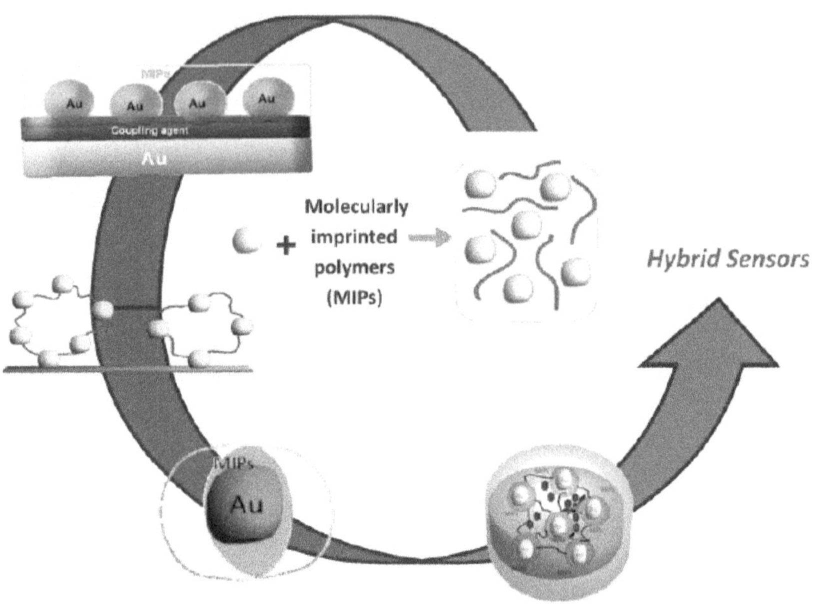

Figure 7.4 Schematic illustration of hybrid synthesis. Reproduced from Reference 76 with permission from American Chemical Society.

methylammonium chloride (MTMA) modifier for biomolecule deposition. For glucose sensing, the substrate was prepared using enzyme glucose oxidase (GOx, β-d-glucose:oxygen-1-oxidoreductase, EC 1.1.3.4). The materials were observed to have good sensing ability towards glucose.

Polyphenazine or polytriphenylmethane redox polymers have been extensively used as biochemical or electrical sensors in the recent years [79]. The synergetic effects as well as the complementary mechanical, electrical and chemical properties of phenazine/triphenylmethane polymers with CNTs lead to enhanced sensing abilities. In a recent study, electrodeposited gold nanoparticles-polyvinylpyrrolidone-polyaniline (AuNp-PVP-PANI) nanocomposites were reported as glucose biosensors by Miao *et al.* [80]. PVP, gold nanoparticles and aniline were mixed to generate a homogeneous solution and were electrodeposited onto glassy carbon electrode (GCE) substrate. Using Nafion as a prevention layer, glucose oxidase (GOx) was successively immobilized onto the electrode material. Nafion/GOx/AuNP-PVP-PANI modified GCE was studied for electrocatalytic and electrochemical properties. The sensitivity of the sensor was observed to be 9.62 µA mM^{-1} cm^{-2}. It also exhibited a lower detection limit of 1.0×10^{-5} M (S/N = 3). The sensor had effective electrocatalytic performance with a linear range from 0.05 mM to 2.25 mM. In addition, the sensor had good reproducibility and stability. It could also be tested in human serum samples with satisfactory results.

Wang *et al.* [81] reported a sensitive and selective aptasensor for the detection of dopamine in human serum samples, based on PEDOT and reduced graphene oxide. The interface in the polymer composite was prepared by electrochemical polymerization of EDOT, employing graphene oxide as the dopant, which electrochemically reduced to form rGO. Covalent functionalization of the composite with aptamer resulted in sensitive and selective sensing. 1 pM to 160 nM concentration range showed a linear response to a voltage of 160 mV. A detection limit of 78 fM was observed, which was reported for the first time. The sensor could be reused after regenerating in urea. In addition, the senor did not respond to the common interferents. The recovery of the samples after applying to human serum samples was 98.3 to 100.7%. In another study, Mazeiko *et al.* [82] developed GOx and gold nanoparticles entrapped within PANI layer (GOx/Au-NPs/PANI) as well as GOx entrapped within PANI layer (GOx/PANI). The amperometric sensing ability of the composites was tested by fabricating them into carbon electrodes. For all evaluated samples, the material modified with gold nanoparticles exhibited better performance owing to their electron transfer mechanism. Overall, GOx/PANI and GOx/Au-NPs/PANI nanocomposites were observed to retain catalytic activity of glucose oxidase, thus, indicating their suitability designing amperometric biosensors.

Dervisevic *et al.* [83] incorporated multi-walled carbon nanotubes (MWCNTs) in poly(GMA-co-VFc) copolymer film for use as a scaffold for immobilizing different enzymes for biosensing. The spoilage of a fish can be detected by the increased level and variability of xanthine. In this respect, the prepared enzyme electrodes exhibited a sensitivity of 16 mAM^{-1}. The composite reached 95% of steady state current in about 4 s. For the detection of xanthine in fish, low detection limit of 0.12 µM, linear ranges (2-28 µM, 28-46 and 46-86 µM) and analytical performance were observed. The composites also exhibited good storage capability and no response towards common interferents. In a recent study, Wang *et al.* [84] also studied the electrochemical catalytic activity of PEDOT doped with GO towards dopamine. Excellent catalytic activity and low electrochemical impedance were observed towards dopamine oxidation. The sensor was observed to be free from common interferents like ascorbic acid and uric acid. It had a linear detection range between 0.1-175 µM with a detection limit of 39 nM.

Electrospraying technique was used by Ruecha *et al.* [85] to modify paper based biosensors using graphene/polyvinylpyrrolidone/polyaniline nanocomposites. The presence of a small amount of PVP enhanced the sensitivity of the biosensor owing to better dispersion of graphene which subsequently enhanced the electrochemical conductivity of the composites. The G/PVP/PANI-modified electrode exhibited a three-fold increase in current signal, as compared to the unmodified electrode. The modified electrode also exhibited better electrocatalytic performance towards oxidation of H_2O_2. For the amperometric detection of cholesterol, the composite could be used to immobilize cholesterol oxidase (ChOx) enzyme. The composite showed 50 µM to 10 mM linear detection range with a 1 µM detection limit. The developed composites have high potential for application in biological fluids for detection of cholesterol. In another study, Njagi *et al.* [86] reported Au-PPy nanocomposites to immobilize different enzymes and use as amperometric biosensors. Cytochrome *c*, glucose oxidase and polyphenol oxidase were used as model enzymes for immobilization. The composite sample was first deposited onto a glassy carbon electrode which was followed by the deposition of enzymes. The composites exhibited good adhesion towards the glassy carbon electrode along with good stability and uniform distribution. The gold nanoparticles present in the composites improved the electrochemical activity. A small amount of pyrrole and enzyme was enough for the biosensor to exhibit optimal performance. The matrix of the composites also prevented the leakage of enzymes. The composites exhibited high sensitivity towards glucose and phenol. Low detection limit and fast response time were the other benefits of the developed sensors.

Sirkar *et al.* [87] employed poly[vinylpyridine Os(bisbipyridine)$_2$Cl]-*co*-allylamine for the fabrication of glucose and lactate enzyme electrodes. For this purpose, the authors carried out the deposition of an anionic self-assembled monolayer and subsequent redox polymer/enzyme electrostatic interaction on gold substrate. It was followed by modification with 11-mercaptoundecanoic acid (MUA), and subsequently alternating immersions in cationic redox polymer solutions and anionic glucose oxidase (GOX) or lactate oxidase (LAX) solutions. Figure 7.5 depicts the multi-layer composite structure generated using this process.

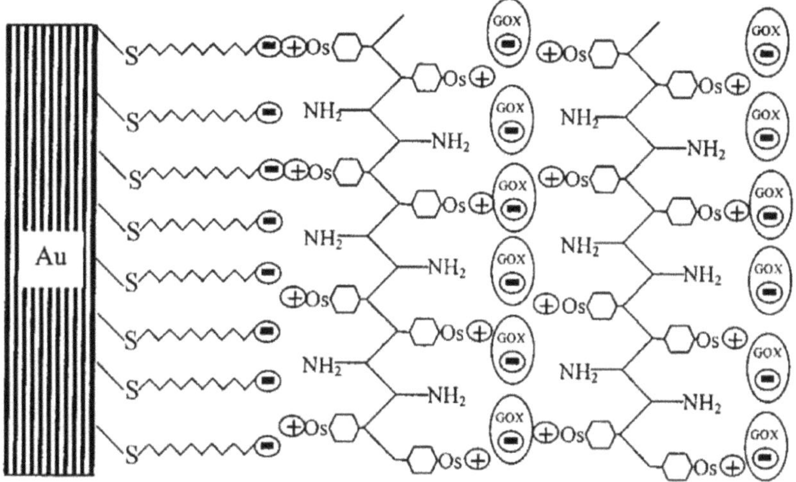

Figure 7.5 Schematics of the multi-layer structure. Reproduced from Reference 87 with permission from American Chemical Society.

7.3.3 Humidity Sensors

Low cost, efficient and high speed microelectronic devices pave the way to develop sensors with practical applications [88-91]. Miniaturization of the devices and modern signal conditioning methods further help in this aspect to achieve functional sensors [92-98]. For better integration of the devices, it is important for these materials to be reliable and specific towards each sensing mechanism. In addition, their calibration technique and principle behind the mechanism of action are important to be known [99]. To improve the quality of the sensors, simulation and design aides have also been widely applied [100-106].

Humidity sensors find a wide range of applications. For instance, in agriculture, understanding the moisture content of soil is important for crops and in civil engineering, the prevention of corrosion and erosion is achieved by taking the humidity into consideration and providing effective prevention techniques [107-109]. Such critical needs have, thus, resulted in the development of humidity sensors which can detect the moisture content even in the presence of different organic and inorganic molecules using various physical and chemical methods of sensing [110-113]. Improvements in design and fabrication, elements of the sensing device like transducers, structure of various devices and mechanism are some methods that have been employed to enhance the performance of humidity sensors [114-125].

Hwang *et al.* [126] reported composites comprising of poly(dopamine)-treated graphene oxide/poly(vinyl alcohol) (dGO/PVA) for use as a humidity sensor. Figure 7.6 presents the schematic of composite synthesis. In general, graphene oxide was treated with dopamine in presence of water. This was followed by the chemical reduction of the material which yielded poly(dopamine)-treated reduced GO. At 0.5 wt % dGO loading, the material exhibited improved tensile modulus, tensile strength and strain to failure. This was achieved by the synergetic effect of strong adhesion of polydopamine between the interface of GO and PVA, hydrogen bonding, etc. The developed material acted as a good humidity sensor over the relative humidity range of 40-100%.

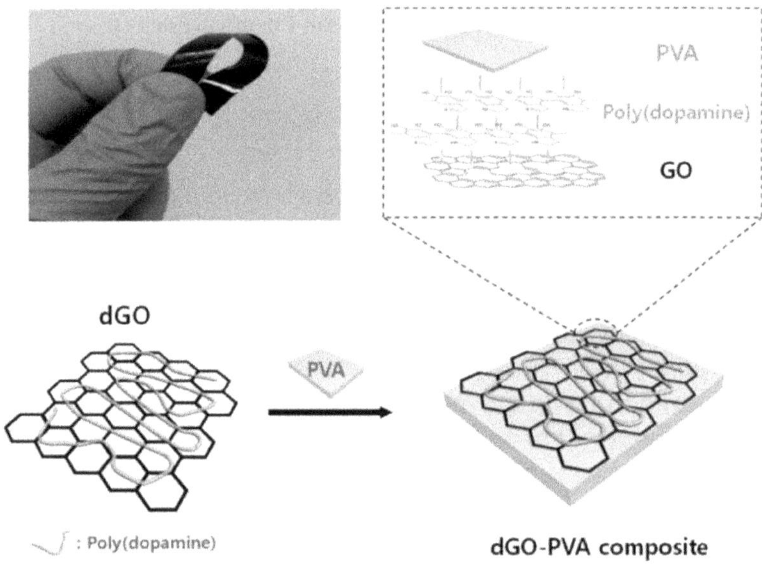

Figure 7.6 Schematics of the structure of a dGO/PVA composite film. Reproduced from Reference 126 with permission from American Chemical Society.

In another study, Zhang *et al.* [127] reported humidity sensor based on GO/poly(diallyldimethylammonium chloride) (PDDA). Layer by layer self-assembly was used for the synthesis of the composite material. The multi-layered film of GO/PDDA was generated on a polyimide substrate. The material exhibited very high response of up to 265,640% when exposed to different humidity ranges. The sensor was also capable of monitoring human health owing to its rapid response and recovery times. Lim *et al.* [128] also reported composites based on PVA/rGO coated with tannic acid (TA) for humidity sensing. The rGO-TA was obtained by simply mixing TA with GO. The simple method of mixing helped in reducing GO and provided a covalent grafting of TA on graphene surface. This also enhanced the affinity of rGO towards PVA which further enhanced the mechanical properties of the polymer composites. With the addition of 1 wt% of rGO-TA, enhancement of 56.9% and 34% in elongation at break and Young's modulus of PVA was observed. The material had a wide range of humidity detection and stability. In another study, Zhang *et al.* [129] also reported humidity sensors by coating tin dioxide/reduced graphene oxide nanocomposite film on a polyimide substrate with microelectrodes. Zhang *et al.* [130] also developed rGO/ PDDA nanocomposite film sensor for humidity sensing applications. The sensor film was generated on a polyimide substrate which had multiple interdigital microelectrodes. Layer by layer self-assembly was used to fabricate the thin film which was followed by solution chemical reduction to reduce the GO. The hierarchical nanostructure had high conductivity and chemically active sites. A wide range of humidity sensing studies were performed on the composite at room temperature to confirm its excellent sensing behavior. Other features which affect the sensing capacity of the sensors like response-recovery characteristics, hysteresis and repeatability were also observed to exhibit promising results.

Sodium polystyrene sulfonate (NaPSS) and ZnO were used to fabricate thin film humidity sensor by Li *et al.* [131]. The composite film sensor was tested for humidity sensing over a range of humidities by measuring complex impedance spectra. For the whole range of humidity, the log of impedance of the sensor changed linearly by four order of magnitude. The sensor had a response time of 2 seconds for both adsorption and desorption. Also, a very low hysteresis was observed for the composite. Compared with the pure NaPSS film, the nanocomposite film demonstrated faster response and improved linearity. Arregui *et al.* [132] also developed an optical fiber based humidity sensor using the ionic self-assembly process. In another study, cellulose was used as a humidity and temperature sensor [133]. *In-situ* polymerization was used to introduce PPy nanoscale layer on the cellulose surface. This process did not damage the structure of the cellulose. Cellulose-PPy nanocomposite generated with 16 h polymerization time was observed to be optimal for humidity and temperature sensing.

Parvatikar *et al.* [134] used ammonium persulphate as an oxidizing agent for generating polyaniline/CeO_2 composites using *in-situ* polymerization. Using this single step polymerization process, emeraldine phase polymer was synthesized. The pressed pellets of the material exhibited a humidity response over a broad range of humidity. High temperature conductivity measurements also exhibited a "thermal activated behavior" for the material. In another study, Su *et al.* [135] synthesized resistive humidity sensors using multi-walled carbon nanotubes, methyl methacrylate (MMA) and KOH composite thin films. The composition and microstructure of the polymer composite played an important role in determining the mechanism and efficiency of humidity sensing. Dimensions of the channels through which ions migrated as well as electrolytic dissociation which caused the charge-transportation by the ions and their interaction with the polymer were used to explain the sensing mechanism of the composite films. Hysteresis, stability, recovery duration, response time as well as linearity were studied and demonstrated promising results. The authors also fabricated thin films of composites comprising PPy and TiO_2 nanoparticles (TiO_2 NPs/PPy) on alumina substrate using *in- situ* photopolymerization technique [136]. The material exhibited resistive humidity sensing properties. Superior linear properties, small hysteresis and high sensitivity were observed for the composite comprising of 0.0012 g TiO_2 NPs. Studies on long term stability, response and recovery time, temperature of operation, applied frequency, etc., confirmed the potential of the sensing material.

7.3.4 Liquid Sensing

The early detection, monitoring and control of the harmful liquids in a variety of industrial effluents is necessary for achieving safe environment [137-144]. In one such study, Potschke *et al.* [145] reported the liquid sensing property of polypropylene (PP)/poly(ε-caprolactone) (PCL) blends containing MWCNTs. Using a solvent immersion and drying cycle, the changes in the electrical resistance of the composites were monitored. For the 50/50 blend composite with 3 wt% loading of CNTs, large detection ability was observed for n-hexane, evident from significant change in resistance. In case of ethanol, PCL-MWNTs composite exhibited better response.

Massaro *et al.* [146] incorporated gold micro/nanoparticles in PDMS for liquid sensing. The authors developed PDMS/PDMS-Au pillar-type layout and monitored the variation in the transmittivity for different liquids, such as ethanol and oil, placed on the PDMS-Au sensor. The authors proposed the use of the developed sensor for robotics finger. Guo *et al.* [147] also used solvent-cast 3D printing to generate a multi-functional 3D liquid sensor using poly(lactic acid) (PLA)/MWCNT nanocomposite. By short immersion in the solvents, the sensor exhibited high sensitivity and selectivity. The helical structure helped in the easier trapping of liquid molecules, which enhanced the sensing efficiency. In another study, Dai *et al.* [148] reported low percolation threshold PP-carbon black composites fabricated by dispersing carbon black on the interfaces between PP polyhedrons. High response rate, low response time and good selectivity were observed for the composites when used as liquid sensors. The interfaces accelerated the migration of solvent molecules by capillary effect.

Kobashi *et al.* [149] studied the changes in electrical properties of the polymer composites fabricated using PLA/MWCNTs on contact with solvents. A very low electrical percolation threshold of 0.5 wt% was observed for the composites. The liquid sensing performance of the composites was facilitated by the conductive network structure of CNTs in PLA. The composites were thin pressed to obtain sheets which were used for solvent immersion and drying cycles. The materials exhibited a reversible change in resistance, along with high reproducibility. Larger resistance changes were observed for lower MWCNTs loadings. This was attributed to the easier disconnection of network structure owing to the less dense network. Water, ethanol, dichloromethane, tetrahydrofuran, chloroform, toluene and *n*-hexane could be detected using the developed sensing materials, though different solvents exhibited different degrees of resistance change.

7.4 Conclusions

A large variety of polymer nanocomposites have been developed and analyzed for sensing applications, most abundant among these being conducting composites. Incorporation of functional reinforcements has been observed to significantly develop sensing attributes of polymers. In this respect, various kinds of sensors like gas sensors, liquid sensors, humidity sensors, biosensors, etc., have been developed for a range of environmental applications.

References

1. Adhikari, B., and Majumdar, S. (2004) Polymers in sensor applications. *Progress in Polymer Science*, **29**(7), 699-766.
2. Stetter, J. R., Penrose, W. R., and Yao, S. (2003) Sensors, chemical sensors, electrochemical sensors, and ECS. *Journal of The Electrochemical Society*, **150**(2) S11-S16.
3. Janata, J., and Bezegh A. (1988) Chemical sensors. *Analytical Chemistry*, **60**, 62-74.
4. Zhu, L., Cao, Y., and Cao, G. (2014) Electrochemical sensor based on magnetic molecularly imprinted nanoparticles at surfactant modified magnetic electrode for determination of bisphenol A. *Biosensors & Bioelectronics*, **54**, 258-261.
5. Long, F., Wang, J., Zhang, Z., and Yan, L. (2016) Magnetic imprinted electrochemical sensor combined with magnetic imprinted solid-phase extraction for rapid and sensitive detection of tetrabromobisphenol S. *Journal of Electroanalytical Chemistry*, **777**, 58-66.
6. McAvoy, D. C., Pittinger, C. A., and Willis A. M. (2016) Biotransformation of tetrabromobisphenol A (TBBPA) in anaerobic digester sludge, soils, and freshwater sediments. *Ecotoxicology and Environmental Safety*, **131**, 143-150.
7. Lachance, B., Renoux, A. Y., Sarrazin, M., Hawari, J., and Sunahara, G. I. (2004) Toxicity and bioaccumulation of reduced TNT metabolites in the earthworm Eisenia andrei exposed to amended forest soil. *Chemosphere*, **55**, 1339-1348.
8. Alizadeh, T. (2014) Preparation of magnetic TNT-imprinted polymer nanoparticles and their accumulation onto magnetic carbon paste electrode for TNT determination. *Biosensors & Bioelectronics*, **61**, 532-540.
9. Song, N. H., Chen, L., and Yang, H. (2008) Effect of dissolved organic matter on mobility and activation of chlorotoluron in soil and wheat. *Geoderma*, **146**, 344-352.
10. Li, X., Zhang, L., Wei, X., and Li, J. (2013) A sensitive and renewable chlortoluron molecularly imprinted polymer sensor based on the gate-controlled catalytic electrooxidation of H_2O_2 on magnetic nano-NiO. *Electroanalysis*, **5**, 1286-1293.
11. Rapini, R., and Marrazza, G. (2016) Biosensor potential in pesticide monitoring. *Comprehensive Analytical Chemistry*, 74, 3-21.
12. Li, Y., Yu, J., and Ding, B. (2015) Facile and ultrasensitive sensors based on electrospinning-netting nanofibers/net. In: Electrospinning High Perform. Sensors, Macagnano, A., Zampetti, E., and Kny. E. (eds.), Springer, Germany, pp. 1-34.
13. Perez-Lopez, B., and Merkoci A. (2011) Nanomaterials based biosensors for food analysis applications. *Trends in Food Science and Technology*, **22**, 625-639.

14. Fenzl, C., Hirsch, T., and Baeumner, A. J. (2016) Nanomaterials as versatile tools for signal amplification in (bio)analytical applications. *TrAC Trends in Analytical Chemistry*, **79**, 306-316.
15. Wang, Y., and Duncan T. V.,(2017) Nanoscale sensors for assuring the safety of food products. *Current Opinion in Biotechnology*, **44**, 74-86.
16. Correa, D. S., Pavinatto, A., Mercante, L. A., Mattoso, L. H. C, Oliveira, J. E, Riul, Jr., A. (2016) Chemical sensors based on hybrid nanomaterials for food analysis. In: *Nanaobiosensors*, Grumezescu, A. H. (ed.), Academic Press, USA, pp. 205-244.
17. Zhu, C., Yang, G., Li, H., Du, D., and Lin, Y. (2015) Electrochemical sensors and biosensors based on nanomaterials and nanostructures. *Analytical Chemistry*, **87**, 230-249.
18. *Electrospinning for High Performance Sensors*, Macagnano, A., Zampetti, E., and Kny, E. (eds.), Springer, Germany (2015).
19. Agarwal, S., Greiner, A., and Wendorff, J. H. (2013) Functional materials by electrospinning of polymers. *Progress in Polymer Science*, **38**, 963-991.
20. Gangopadhyay, R., and De, A. (2000) Conducting polymer nanocomposites: A brief overview. *Chemistry of Materials*, **12**(3), 608-622.
21. Walker, C. H., St. John, J. V., and Wisian-Neilson, P. (2001) Synthesis and size control of gold nanoparticles stabilized by poly (methylphenyl phosphazene). *Journal of American Chemical Society*, **123**, 3846-3847.
22. Wiley, B., Herricks, T., Sun, Y., and Xia, Y. (2004) Polyol synthesis of silver nanoparticles: use of chloride and oxygen to promote the formation of single-crystal, truncated cubes and tetrahedrons. *Nano Letters*, **4**, 1733-1739.
23. Iijima, S. (1991) Helical microtubules of graphite carbon. *Nature*, **354**, 56-58.
24. Nath, M., and Rao, C. N. R. (2001) New metal disulfide nanotubes. *Journal of American Chemical Society*, **123**, 4841-4842.
25. Qiu, H., Zhai, J., Li, S., Jiang, L., and Wan M. (2003) Oriented growth of self-assembled polyaniline nanowires arrays using a novel method. *Advanced Functional Materials*, **13**, 925-928.
26. Walter, E. C., Zach, M. P., Favier F., Murray, B. J., Inazu, K., Hemminger, J. C., and Penner, R. M. (2003) Metal nanowires arrays by electrodeposition. *ChemPhysChem*, **4**, 131-138.
27. Liang, L., Liu, J., Windisch, Jr., C. F., Exarhos, G. J., and Lin, Y. (2002) Direct assembly of large arrays of oriented conducting polymer nanowires. *Angewandte Chemie, International Edition*, **41**, 3665-3668.
28. *Encyclopedia of Nanoscience and Nanotechnology*, Nalwa, H. S. (ed.), American Scientific Publishers, USA (2003).
29. Cui, Y., Wei, Q., Park, H., and Lieber, C. M. (2001) Nanowire nanosensors for highly sensitive and selective detection of biological

and chemical species. *Science*, **293**, 1289-1292.

30. Favier, F., Walter, E. C, Zach, M. P., Benter, T., and Penner, R. M. (2001) Hydrogen sensors and switches from electrodeposited palladium mesowire arrays. *Science*, **293**, 2227-2231.

31. Wang, J. (2005) Nanomaterial-based electrochemical biosensors. *Analyst*, **130**, 421-426.

32. Jain, K. K. (2005) Nanotechnology in clinical laboratory diagnostics. *Clinica Chimica Acta*, **358**, 37-54.

33. Wu, X., Lu, C., Han, Y., Zhou, Z., Yuan, G., and Zhang, X. (2016) Cellulose nanowhisker modulated 3D hierarchical conductive structure of carbon black/natural rubber nanocomposites for liquid and strain sensing application, *Composites Science and Technology*, **124**, 44-51.

34. Sherigara, B. S., Kutner, W. and D'Souza, F. (2003) Electrocatalytic properties and sensor applications of fullerenes and carbon nanotubes. *Electroanalysis*, **15**, 753-772.

35. Wang, J. (2005) Carbon-nanotube based electrochemical biosensors: A review. *Electroanalysis*, **17**, 7-14.

36. Hrapovic, S., Liu, Y., Male, K. B., and Luong, J. H. T (2004) Electrochemical biosensing platforms using Pt nanoparticles and carbon nanotubes. *Analytical Chemistry*, **76**,1083-1088.

37. Zou, Y., Xiang, C., Sun, L.-X., and Xu, F. (2008) Glucose biosensor based on electrodeposition of platinum nanoparticles onto carbon nanotubes and immobilizing enzyme with chitosan-SiO_2 sol gel. *Biosensors & Bioelectronics*, **23**, 1010-1016.

38. Ramanavicius, A., Kausaite, A., and Ramanaviciene, A. (2005) Polypyrrole-coated glucose oxidase nanoparticles for biosensor design. *Sensors and Actuators B*, **111-112**, 532-539.

39. Somasundrum, M., Kirtikara, K., and Tanticharoen, M. (1996) Amperometric determination of hydrogenperoxide by direct and catalytic reduction at copper electrode. *Analytica Chimica Acta*, **319**, 59-70.

40. Willner, I., Baron, R., and Willner, B. (2006) Integrated nanoparticles-biomolecule systems for biosensing and bioelectronics. *Biosensors & Bioelectronics*, **22**, 1841-1852.

41. Pradhan, B., Setyowati, K., Liu, H., Waldeck, D. H., and Chen, J. (2008) Carbon nanotube-polymer nanocomposite infrared sensor. *Nano Letters*, **8**, 1142-1146.

42. Macanas, J., Farre, M., Munoz ,M., Alegret, S., and Muraviev, D. N. (2006) Preparation and characterization of polymer-stabilized metal nanoparticles for sensor applications. *Physica Status Solidi A*, **203**, 1194-1200.

43. Shirikawa, H., Louis, E. J, MacDiarmid, A. G., Chiang, C. K., and Heeger, A. J. (1977) Synthesis of electrically conducting organic polymers: halogen derivatives of polyacetylene, (CH)ₓ, *Journal of the Chemical*

Society, Chemical Communications, 578-580.

44. *Handbook of Conducting Polymers*, Skotheim, T. A., and Reynolds, J. (eds.), 3rd edition, CRC Press, USA (2007).

45. Cichosz, S., Masek, A., and Zaborski, M. (2018) Polymer-based sensors: A review. *Polymer Testing*, **67**, 342-248.

46. Retama J. R., Mecerreyes, D., Lopez-Ruiz, B., and Lopez-Cabarcos, E. (2005) Synthesis and characterization of semi conducting polypyrrole/polyacrylamide micro particles with GOx for biosensor application. *Colloids and Surfaces A: Physicochemical and Engineering Aspects*, **270-271**, 239-244.

47. Lange, U., Roznyatovskaya, N. V, and Mirsky V. M. (2008) Conducting polymers in chemical sensors and arrays. *Analytica Chimica Acta*, **614**, 1-26.

48. Gupta, N., Sharma, S., Mir, I. A., and Kumar, D. (2006) Advances in sensors based on conducting polymers. *Journal of Scientific & Industrial Research*, **65**, 549-557.

49. Cosnier, S. (2005) Affinity biosensors based on electropolymerized films. *Electroanalysis*, **17**, 1701-1715.

50. Al Sana, S., Gabrielli, C., and Perrot, H. (2003) Influence of antibody insertion on the electrochemical behavior of polypyrrole films by using fast QCM measurements. *Journal of the Electrochemical Society*, **150**, E444-E449.

51. Kim, B. H., Kim, M. S., Park, K. T., Lee, J. K., Park, D. H., Joo, J., Yu, S. G., and Lee, S. H. (2003) Characteristics and field emission of conducting poly (3,4-ethylenedioxythiophene) nanowires. *Applied Physics Letters*, **83**, 539-541.

52. Huang, J. (2006) Syntheses and applications of conducting polymer polyaniline nanofibers. *Pure and Applied Chemistry*, **78**, 15-27.

53. Hatchett, D. W., and Josowicz, M. (2008) Composites of intrinsically conducting polymers as sensing nanomaterials. *Chemical Reviews*, **108**, 746-769.

54. Malinauskas, A., Malinauskiene, J., and Ramanavicius, A. (2005) Conducting polymer-based nanostructurized materials: Electrochemical aspects. *Nanotechnology*, **16**, R51-R62.

55. Rajesh, Ahuja, T., and Kumar, D. (2009) Recent progress in the development of nano-structured conducting polymers/nanocomposites for sensor applications. *Sensors and Actuators B: Chemical*, **136**, 275-286.

56. Ram, M. K., Yavuz, O., and Aldissi, M. (2005) NO gas sensing based on ordered ultrathin films of conducting polymer and its nanocomposite. *Synthetic Metals*, **151**, 77-84.

57. Ram, M. K., Yavuz, O., Lahsangah, V., and Aldissi, M. (2005) CO gas sensing from ultrathin nano-composite conducting polymer film. *Sensors and Actuators B: Chemical*, **106**, 750-757.

58. Amjadi, M., Pichitpajongkit, A., Lee, S., Ryu, S., and Park, I. (2014)

Highly stretchable and sensitive strain sensor based on silver nanowire-elastomer nanocomposite. *ACS Nano*, **8**(5), 5154-5163.

59. Kaushik, A., Kumar, R., Arya, S. K., Nair, M., Malhotra, B. D. and Bhansali, S. (2015) Organic-inorganic hybrid nanocomposite-based gas sensors for environmental monitoring. *Chemical Reviews*, **115**(11), 4571-4606.

60. Abraham, J. K., Philip, B., Witchurch, A., Varadan, V. K., and Reddy, C. C. (2004) A compact wireless gas sensor using a carbon nanotube/PMMA thin film chemiresistor. *Smart Materials and Structures*, **13**, 1045.

61. Tai, H., Jiang, Y., Xie, G., Yu, J., and Chen, X. (2007) Fabrication and gas sensitivity of polyaniline–titanium dioxide nanocomposite thin film. *Sensors and Actuators B: Chemical*, **125**, 644-650.

62. Tai, H., Jiang, Y., Xie, G., Yu, J., Chen, X., and Ying, Z. (2008) Influence of polymerization temperature on NH_3 response of PANI/TiO thin film gas sensor. *Sensors and Actuators B: Chemical*, **129**, 319-326.

63. Suri, K., Annapoorni, S., Sarkar, A. K, Tandon, R. P. (2002) Gas and humidity sensors based on iron oxide-polypyrrole nanocomposites. *Sensors and Actuators B: Chemical*, **81**, 277-282.

64. Mishra, S. K., Tripathi, S. N., Choudhary, V., and Gupta, B. D. (2014) SPR based fibre optic ammonia gas sensor utilizing nanocomposite film of PMMA/reduced graphene oxide prepared by in situ polymerization. *Sensors and Actuators B: Chemical*, **199**, 190-200.

65. Chen, J., and Tsubokawa, N. (2000) A novel gas sensor from polymer-grafted carbon black: responsiveness of electric resistance of conducting composite from LDPE and PE-b-PEO-grafted carbon black in various vapors. *Polymers for Advanced Technologies*, **11**, 101-107.

66. Shang, S., Li, L., Yang, X., and Wei, Y. (2009) Polymethylmethacrylate-carbon nanotubes composites prepared by microemulsion polymerization for gas sensor. *Composites Science and Technology*, **69**, 1156-1159.

67. Tandon, R. P., Tripathy, M. R., Arora, A. K., and Hotchandani, S. (2006) Gas and humidity response of iron oxide-polypyrrole nanocomposites. *Sensors and Actuators B: Chemical*, **114**, 768-773.

68. Yang, Y., Li. S., Yang, W., Yuan, W., Xu, J., and Jiang, Y. (2014) In situ polymerization deposition of porous conducting polymer on reduced graphene oxide for gas sensor. *ACS Applied Materials & Interfaces*, **6**, 13807-13814.

69. Sadek, A. Z., Wlodarski, W., Shin, K., Kaner, R. B. and Kalantar-zadeh, K. (2006) A layered surface acoustic wave gas sensor based on a polyaniline/In_2O_3 nanofibre composite. *Nanotechnology*, **17**, 4488.

70. Karadas, N., and Ozkan, S. A. (2014) Electrochemical preparation of sodium dodecylsulfate doped over-oxidized polypyrrole/multi-walled carbon nanotube composite on glassy carbon electrode and

its application on sensitive and selective determination of anticancer drug: Pemetrexed. *Talanta*, **119**, 248-254.

71. Shahrokhian, S., Azimzadeh, M., and Amini, M. K. (2015) Modification of glassy carbon electrode with a bilayer of multiwalled carbon nanotube/tiron-doped polypyrrole: Application to sensitive voltammetric determination of acyclovir. *Materials Science & Engineering C*, **53**, 134-141.

72. Yang, Y., Fang, G., Wang, X., Pan, M., Qian, H., Liu, H., and Wang S. (2014) Sensitive and selective electrochemical determination of quinoxaline-2-carboxylic acid based on bilayer of novel poly(pyrrole) functional composite using one-step electropolymerization and molecularly imprinted poly(o-phenylenediamine). *Analytica Chimica Acta*, **806**, 136-143.

73. Xue, C., Wang, X., Zhu, W., Han, Q., Zhu, C., Hong, J., Zhou, X., and Jiang, H. (2014) Electrochemical serotonin sensing interface based on double-layered membrane of reduced graphene oxide/polyaniline nanocomposites and molecularly imprinted polymers embedded with gold nanoparticles. *Sensors and Actuators B: Chemical*, **196**, 57-63.

74. Shrivastava, S., Jadon, N., and Jain, R. (2016) Next generation polymer nanocomposites based electrochemical sensors and biosensors: A review. *TrAC Trends in Analytical Chemistry*, **82**, 55-67.

75. Ramanavicius, A., Oztekin, Y., Balevicius, Z., Kausaite-Mikstimiene, A., Krikstolaityte, V., Baleviciute, I., Ratautaite, V., and Ramanaviciene, A. (2012) Conducting and electrochemically generated polymers in sensor design (mini review). *Procedia Engineering*, **47**, 825-828.

76. Ahmad, R., Griffete, N., Lamouri, A., Felidj, N., Chehimi, M. M., and Mangeney, C. (2015) Nanocomposites of gold nanoparticles@molecularly imprinted polymers: Chemistry, processing, and applications in sensors. *Chemistry of Materials*, **27**, 5464-5478.

77. Baghayeri, M., Nazarzadeh Zare, E., and Mansour Lakouraj, M. (2014) A simple hydrogen peroxide biosensor based on a novel electro-magnetic poly(p-phenylenediamine)@Fe_3O_4 nanocomposite. *Biosensors and Bioelectronics*, **55**, 259-265.

78. Emre, F. B., Kesik, M., Kanik, F. E., Akpinar, H. Z., Aslan-Gurel, E., Rossi, R. M., and Toppare, L. (2015) A benzimidazole-based conducting polymer and a PMMA–clay nanocomposite containing biosensor platform for glucose sensing. *Synthetic Metals*, **207**, 102-109.

79. Barsan, M. M., Ghica, M. E., and Brett, C. M. A. (2015) Electrochemical sensors and biosensors based on redox polymer/carbon nanotube modified electrodes: A review. *Analytica Chimica Acta*, **881**, 1-23.

80. Miao, Z., Wang, P., Zhong, A.-M., Yang, M., Xu, Q., Hao, S., and Hu, X. (2015) Development of a glucose biosensor based on electrodeposited gold nanoparticles–polyvinylpyrrolidone–polyaniline nanoco-

mposites. *Journal of Electroanalytical Chemistry*, **756**, 153-160.

81. Wang, W., Wang, W., Davis, J. J., and Luo, X. (2015) Ultrasensitive and selective voltammetric aptasensor for dopamine based on a conducting polymer nanocomposite doped with graphene oxide. *Microchimica Acta*, **182**, 1123-1129.

82. Mazeiko, V., Kausaite-Minkstimiene, A., Ramanaviciene, A., Balevicius, Z., and Ramanavicius, A. (2013) Gold nanoparticle and conducting polymer-polyaniline-based nanocomposites for glucose biosensor design. *Sensors and Actuators B: Chemical*, **189**, 187-193.

83. Dervisevic, M., Custiuc, E., Cevik, E., and Senel, M. (2015) Construction of novel xanthine biosensor by using polymeric mediator/MWCNT nanocomposite layer for fish freshness detection. *Food Chemistry*, **181**, 277-283.

84. Wang, W., Xu, G., Cui, X. T., Sheng, G., and Luo, X. (2014) Enhanced catalytic and dopamine sensing properties of electrochemically reduced conducting polymer nanocomposite doped with pure graphene oxide. *Biosensors and Bioelectronics*, **58**, 153-156.

85. Ruecha, N., Rangkupan, R., Rodthongkum, N., and Chailapakul, O. (2014) Novel paper-based cholesterol biosensor using graphene/polyvinylpyrrolidone/polyaniline nanocomposite. *Biosensors and Bioelectronics*, **52**, 13-19.

86. Njagi, J., and Andreescu, S. (2007) Stable enzyme biosensors based on chemically synthesized Au–polypyrrole nanocomposites. *Biosensors and Bioelectronics*, **23**, 168-175.

87. Sirkar, K., Revzin, A., and Pishko, M. V. (2000) Glucose and lactate biosensors based on redox polymer/oxidoreductase nanocomposite thin films. *Analytical Chemistry*, **72**, 2930-2936.

88. Dai, C.-L., Liu, M.-C., Chen, F.-S., Wu, C.-C., and Chang, M.-W. (2007) A nanowire WO_3 humidity sensor integrated with micro-heater and inverting amplifier circuit on chip manufactured using CMOS-MEMS technique. *Sensors and Actuators B: Chemical*, **123**, 896-901.

89. Boltshauser, T., Schonholzer, M., Brand, O., and Baltes, H. (1992) Resonant humidity sensors using industrial CMOS-technology combined with postprocessing. *Journal of Micromechanics and Microengineering*, **2**, 205.

90. Okcan, B., and Akin, T. A (2007) Low-power robust humidity sensor in a standard CMOS process. *IEEE Transactions on Electron Devices*, **54**, 3071-3078.

91. Oprea, A., Courbat, J., Barsan, N., Briand, D., de Rooij, N. F., and Weimar, U. (2009) Temperature, humidity and gas sensors integrated on plastic foil for low power applications. *Sensors and Actuators B: Chemical*, **140**, 227-232.

92. Shi, Y., Luo, Y., Zhao, W., Shang, C., Wang, Y., and Chen, Y. (2013) A radiosonde using a humidity sensor array with a platinum resistance heater and multi-sensor data fusion. *Sensors*, **13**, 8977-

8996.

93. Christian, S. (2002) New generation of humidity sensors. *Sensor Review*, **22**, 300-302.

94. Mehrabani, S., Kwong, P., Gupta, M., and Armani, A. M. (2013) Hybrid microcavity humidity sensor. *Applied Physics Letters*, **102**, 241101.

95. Wang, Y.-H., Lee, C.-Y., and Chiang, C.-M. (2007) A MEMS-based air flow sensor with a free-standing micro-cantilever structure. *Sensors*, **7**, 2389-2401.

96. Su, P.-G., Ho, C.-J., Sun, Y.-L., and Chen, I.-C. (2006) A micromachined resistive-type humidity sensor with a composite material as sensitive film. *Sensors and Actuators B: Chemical*, **113**, 837-842.

97. Hanreich, G., Nicolics, J., Mundlein, M., Hauser, H., and Chabicovsky, R. (2001) A new bonding technique for human skin humidity sensors. *Sensors and Actuators A: Physical*, **92**, 364-369.

98. Mohd Syaifudin, A. R., Mukhopadhyay, S. C., and Yu, P. L. (2012) Modelling and fabrication of optimum structure of novel interdigital sensors for food inspection. *International Journal of Numerical Modelling, Electronic Networks, Devices and Fields*, **25**, 64-81.

99. Trankler, H.-R., and Kanoun, O. (2001) Recent Advances in Sensor Technology. *Proceedings of the 18th IEEE Instrumentation and Measurement Technology Conference*, Hungary, doi: 10.1109/IMTC.2001.928831.

100. Tetelin, A., and Pellet, C. (2006) Modeling and optimization of a fast response capacitive humidity sensor. *IEEE Sensors Journal*, **6**, 714-720.

101. Suchanek, P., and Husak, M. (2006) Design and Simulation of Humidity Micro-Sensors Structure Based on Polymers. *2006 International Conference on Advanced Semiconductor Devices and Microsystems*, Slovakia, doi:10.1109/ASDAM.2006.331186.

102. Connolly, E. J., O'Halloran, G. M., Pham, H. T. M., Sarro, P. M., and French, P. J. (2002) Comparison of porous silicon, porous polysilicon and porous silicon carbide as materials for humidity sensing applications. Sensors and Actuators A: Physical, 99, 25-30.

103. Goldberg, H. D., Brown, R. B., Liu, D. P., and Meyerhoff, M. E. (1994) Screen printing: A technology for the batch fabrication of integrated chemical-sensor arrays. *Sensors and Actuators B: Chemical*, **21**, 171-183.

104. Krutovertsev, S. A., Tarasova, A. E., Krutovertseva, L. S., Chuprin, M. V., Ivanova, O. M., and Sazhinev, Y. S. (2011) Technology and Characteristics of Microhumidity Sensors. *2011 16th International Solid-State Sensors, Actuators and Microsystems Conference*, China, doi:10.1109/TRANSDUCERS.2011.5969782.

105. Lee, C.-Y., and Lee, G.-B. (2003) Micromachine-based humidity sensors with integrated temperature sensors for signal drift compensation. *Journal of Micromechanics and Microengineering*, **13**, 620.

106. Kandler, M., Manoli, Y., Mokwa, W., Spiegel, E., and Vogt, H. (1992) A miniature single-chip pressure and temperature sensor. *Journal of Micromechanics and Microengineering*, **2**, 199.

107. Yang, B., Aksak, B., Lin, Q., and Sitti, M. (2006) Compliant and low-cost humidity nanosensors using nanoporous polymer membranes. *Sensors and Actuators B: Chemical*, **114**, 254-262.

108. Eyebe, G. A., Bideau, B., Boubekeur, N., Loranger, E., and Domingue, F. (2017) Environmentally-friendly cellulose nanofibre sheets for humidity sensing in microwave frequencies. *Sensors and Actuators B: Chemical*, **245**, 484-492.

109. Carr-Brion, K. (1986) *Moisture Sensors in Process Control*, Elsevier Applied Science Publishers, UK.

110. Dean, R. N., Rane, A. K., Baginski, M. E., Richard, J., Hartzog, Z., and Elton, D. J. (2012) A capacitive fringing field sensor design for moisture measurement based on printed circuit board technology. *IEEE Transactions on Instrumentation and Measurement*, **61**, 1105-1112.

111. Salehi, A., Nikfarjam, A., and Kalantari, D. J. (2006) Highly sensitive humidity sensor using Pd/porous GaAs Schottky contact. *IEEE Sensors Journal*, **6**, 1415-1421.

112. Kim, J.-H., Hong, S.-M., Lee, J.-S., Moon, B.-M., and Kim, K. (2009) High sensitivity capacitive humidity sensor with a novel polyimide design fabricated by MEMS technology. *2009 4th IEEE International Conference on Nano/Micro Engineered and Molecular Systems*, China, doi:10.1109/NEMS.2009.5068676.

113. Xu, L., Wang, R., Xiao, Q., Zhang, D., and Liu, Y. (2011) Micro humidity sensor with high sensitivity and quick response/recovery based on ZnO/TiO_2 composite nanofibers. *Chinese Physics Letters*, **28**, 070702.

114. Aziz, F., Sayyad, M. H., Sulaiman, K., Majlis, B. H., Karimov, K. S., Ahmad, Z., and Sugandi, G. (2012) Corrigendum: Influence of humidity conditions on the capacitive and resistive response of an Al/VOPc/Pt co-planar humidity sensor. *Measurement Science and Technology*, **23**, 069501.

115. Lin, W.-D., Chang, H.-M., and Wu, R.-J. (2013) Applied novel sensing material graphene/polypyrrole for humidity sensor. *Sensors and Actuators B: Chemical*, **181**, 326-331.

116. Yadav, B. C., and Singh, M. (2010) Morphological and humidity sensing investigations on niobium, neodymium, and lanthanum oxides. *IEEE Sensors Journal*, **10**, 1759-1766.

117. Xu, C.-N., Miyazaki, K., and Watanabe, T. (1998) Humidity sensors using manganese oxides. *Sensors and Actuators B: Chemical*, **46**, 87-96.

118. Pelino, M., Colella, C., Cantalini, C., Faccio, M., Ferri, G., and D'Amico, A. (1992) Microstructure and electrical properties of an α-hematite ceramic humidity sensor. *Sensors and Actuators B: Chemical*, **7**, 464-

469.

119. Klym, H., Hadzaman, I., Shpotyuk, O., and Brunner, M. (2011) Multifunctional T/RH-Sensitive Thick-Film Structures for Environmental Sensors. *SENSOR + TEST Conferences 2011*, Germany, pp. 744-748.

120. Traversa, E., and Bearzotti, A. (1995) A novel humidity-detection mechanism for ZnO dense pellets. *Sensors and Actuators B: Chemical*, **23**, 181-186.

121. Gusmano, G., Montesperelli, G., Nunziante, P., and Traversa, E. (1993) Study of the conduction mechanism of $MgAl_2O_4$ at different environmental humidities. *Electrochimica Acta*, **38**, 2617-2621.

122. Morten, B., Prudenziati, M., and Taroni, A. (1983) Thick-film technology and sensors. *Sensors and Actuators*, **4**, 237-245.

123. Smetana, W., and Unger, M. (2007) Design and characterization of a humidity sensor realized in LTCC-technology. *Microsystem Technologies*, **14**, 979-987.

124. Karimov, K. S., Cheong, K. Y., Saleem, M., Murtaza, I., Farooq, M., and Noor, A. F. M. (2010) Ag/PEPC/NiPc/ZnO/Ag thin film capacitive and resistive humidity sensors. *Journal of Semiconductors*, **31**, 054002.

125. Farahani, H., Wagiran, R., and Hamidon, M. N. (2014) Humidity sensors principle, mechanism, and fabrication technologies: a comprehensive review. *Sensors*, **14**(5), 7881-7939.

126. Hwang, S.-H., Kang, D., Ruoff, R. S., Shin, H. S., and Park, Y.-B. (2014) Poly(vinyl alcohol) reinforced and toughened with poly(dopamine)-treated graphene oxide, and its use for humidity sensing. *ACS Nano*, **8**, 6739-6747.

127. Zhang, D., Tong, J., Xia, B., and Xue, Q. (2014) Ultrahigh performance humidity sensor based on layer-by-layer self-assembly of graphene oxide/polyelectrolyte nanocomposite film. *Sensors and Actuators B: Chemical*, **203**, 263-270.

128. Lim, M.-Y., Shin, H., Shin, D. M., Lee, S.-S., and Lee, J.-C. (2016) Poly(vinyl alcohol) nanocomposites containing reduced graphene oxide coated with tannic acid for humidity sensor. *Polymer*, **84**, 89-98.

129. Zhang, D., Chang, H., Li. P., Liu, R., and Xue, Q. (2016) Fabrication and characterization of an ultrasensitive humidity sensor based on metal oxide/graphene hybrid nanocomposite. *Sensors and Actuators B: Chemical*, **225**, 233-240.

130. Zhang, D., Tong, J., and Xia, B. (2014) Humidity-sensing properties of chemically reduced graphene oxide/polymer nanocomposite film sensor based on layer-by-layer nano self-assembly. *Sensors and Actuators B: Chemical*, **197**, 66-72.

131. Li, Y., Yang, M. J., and She Y. (2004) Humidity sensors using in situ synthesized sodium polystyrenesulfonate/ZnO nanocomposites. *Talanta*, **62**, 707-712.

132. Arregui, F. J., Liu, Y., Matias, I. R., and Claus, R. O. (1999) Optical fiber humidity sensor using a nano Fabry-Perot cavity formed by the ionic self-assembly method. *Sensors and Actuators B: Chemical*, **59**, 54-59.

133. Mahadeva, S. K., Yun, S., and Kim, J. (2011) Flexible humidity and temperature sensor based on cellulose-polypyrrole nanocomposite. *Sensors and Actuators A: Physical*, **165**, 194-199.

134. Parvatikar, N., Jain, S., Bhoraskar, S. V., and Ambika Prasad, M. V. N. (2006) Spectroscopic and electrical properties of polyaniline/CeO$_2$ composites and their application as humidity sensor. *Journal of Applied Polymer Science*, **102**, 5533-5537.

135. Su. P.-G., and Wang, C.-S. (2007) In situ synthesized composite thin films of MWCNTs/PMMA doped with KOH as a resistive humidity sensor. *Sensors and Actuators B: Chemical*, **124**, 303-308.

136. Su, P.-G., and Huang, L.-N. (2007) Humidity sensors based on TiO$_2$ nanoparticles/polypyrrole composite thin films. *Sensors and Actuators B: Chemical*, **123**, 501-507.

137. Rentenberger, R., Cayla, A., Villmow, T., Jehnichen, D., Campagne, C., Rochery, M., Devaux, E., and Potschke, P. (2011) Multifilament fibres of poly(ε-caprolactone)/poly(lactic acid) blends with multiwalled carbon nanotubes as sensor materials for ethyl acetate and acetone. *Sensors and Actuators B: Chemical*, **160**, 22-31.

138. Kobashi, K., Villmow, T., Andres, T., Haussler, L., and Potschke, P. (2009) Investigation of liquid sensing mechanism of poly(lactic acid)/multi-walled carbon nanotube composite films. *Smart Materials and Structures*, **18**, 035008.

139. Acikbas, Y., Capan, R., Erdogan, M., and Yukruk, F. (2011) Thin film characterization and vapor sensing properties of a novel perylenediimide material. *Sensors and Actuators B: Chemical*, **160**, 65-71.

140. Segal, E., Tchoudakov, R., Narkis, M., and Siegmann, A. (2003) Sensing of liquids by electrically conductive immiscible polypropylene/thermoplastic polyurethane blends containing carbon black. *Journal of Polymer Science, Part B: Polymer Physics*, **41**(12), 1428-1440.

141. Narkis, M., Srivastava, S., Tchoudakov, R., and Breuer, O. (2000) Sensors for liquids based on conductive immiscible polymer blends. *Synthetic Metals*, **113**, 29-34.

142. Srivastava, S., Tchoudakov, R., and Narkis, M. (2000) A preliminary investigation of conductive immiscible polymer blends as sensor materials. *Polymer Engineering and Science*, **40**(7), 1522-1528.

143. Segal, E., Tchoudakov, R., Narkis, M., and Siegmann, A. (2002) Thermoplastic polyurethane-carbon black compounds: Structure, electrical conductivity and sensing of liquids. *Polymer Engineering and Science*, **42**(12), 2430-2439.

144. Segal, E., Tchoudakov, R., Mironi-Harpaz, I., Narkis, M., and Siegma-

nn, A. (2005) Chemical sensing materials based on electrically-conductive immiscible polymer blends. *Polymer International*, **54**(7), 1065-1075.

145. Potschke, P., Kobashi, K., Villmow, T., Andres, T., Paiva, M. C., and Covas, J. A. (2011) Liquid sensing properties of melt processed polypropylene/poly(ε-caprolactone) blends containing multiwalled carbon nanotubes. *Composites Science and Technology*, **71**, 1451-1460.

146. Massaro, A., Spano, F., Cingolani, R., and Athanassiou A. (2011) Experimental optical characterization and polymeric layouts of gold PDMS nanocomposite sensor for liquid detection. *IEEE Sensors Journal*, **11**, 1780-1786.

147. Guo, S.-z., Yang, X., Heuzey, M.-C., and Therriault, D. (2015) 3D printing of a multifunctional nanocomposite helical liquid sensor. *Nanoscale*, **7**, 6451-6456.

148. Dai, K., Zhao, S., Zhai, W., Zheng, G., Liu, C., Chen, J., and Shen, C. (2013) Tuning of liquid sensing performance of conductive carbon black (CB)/polypropylene (PP) composite utilizing a segregated structure. *Composites, Part A: Applied Science and Manufacturing*, **55**, 11-18.

149. Kobashi, K., Villmow, T., Andres, T., and Pptschke, P. (2008) Liquid sensing of melt-processed poly(lactic acid)/multi-walled carbon nanotube composite films. *Sensors and Actuators B: Chemical*, **134**(2), 787-795.

8

Liquid Crystalline Polymer Composites

8.1 Introduction

Properties of the liquid crystals (LCs) and polymers are synergistically combined in the liquid crystalline polymers (LCPs) [1,2]. Novel high performance polymeric materials have been developed using the anisotropy present in the liquid crystalline mesophase. The liquid crystalline polymers exhibit significant ability to form high strength fibers [3-6]. In addition, the aromatic liquid crystalline polymers have been used in many advanced products, formed using injection molding. Binary and ternary blends of these polymers have also been developed [7]. The magnetic properties of the liquid crystalline polymers consisting of lyotropic or thermotropic LCs have also been utilized for different uses. The isotropic LCs included DNA, polyribonucleotides and polypeptides, whereas the thermotropic LCs comprised polypeptides and polyesters. Numerous studies have reported the development of LCPs using different polymerization techniques [8-26]. The ease of processing and optimal properties also make them attractive materials for developing nanocomposites with fillers like titanium dioxide (TiO_2), silicates, montmorillonite (MMT), carbon nanotubes (CNTs), graphene, etc.

8.2 Liquid Crystals

The common characteristics of the LCs include polarizability, strong dipoles, rigidity of the long axis, etc. [27-28]. One of the most distinguishable property of the liquid crystals is the ability to direct towards a common axis, termed as director. The molecules in a liquid crystal are generally referred to as mesogens. The molecular order in the liquid crystals is different from the order of molecules in the liquid phase, where no particular order is demonstrated. On the other hand, in the solid state, there is no free molecular movement, and the molecules are highly ordered. The mesogenic phase in the liquid crystals, thus, exhibits a molecular arrangement in between the liquid and

Liyamol Jacob and Vikas Mittal, Khalifa University of Science and Technology, Abu Dhabi, UAE

solid phases [29-31]. In terms of order parameters, the liquid crystals exhibit a very high order compared to the isotropic liquids. However, they still demonstrate a high degree of freedom in terms of molecular oscillation, translation, rotation and conformational changes, among others [32-35].

8.3 Classification of LCPs

The LCPs are mainly classified as the main chain LCPs and side chain LCPs, with differences in properties. These are outlined in the following sections.

8.3.1 Main Chain LCPs

A flexible spacer helps in linking the mesogenic groups in the polymer scaffold, and the resulting materials are called the main chain LCPs. The mesophase structure is generally referred to as the low molecular weight structure. The polymer structure must be compatible enough to support the structure of the mesogenic phase. This results in a strong coupling between the mesogenic and polymeric properties. A large number of main chain LCPs like poly(azomethine-ester)s, poly(azomethine)s, poly(esterimide)s, polyimides, polyethers, polycarbonates, polyurethanes, poly(p-phenylene)s, polyamides and polyesters, with a range of different properties, have been widely studied [36-54].

8.3.2 Side Chain LCPs

The side chain LCPs consist of mesogenic side chains, thus, avoiding the use of methylene spacers [55-62]. Success from the earlier studies specifically motivated the development of the side chains based on methacrylates and the backbones consisting of the acrylate/methacrylate chains [63-74]. An orientational order is achieved by packing the mesogens in a parallel order. Compared to the main chain LCPs, the orientation occurs faster in the side chain LCPs. Owing to this, the use of electric field leads to a significant and swift change in the optical properties of the side chain polymers.

At temperatures greater than the glass transition temperature, the anisotropic ordering of the main and side chains is prevented due to the exceptional mobility shown by the chain segments in the liquid phase. In case a flexible spacer has been used to connect the main chain and mesogenic groups, a decoupling in the motion of the main and side chains may take place. In such cases, the conformation in the main chain results in the inhibition of the anisotropic ordering even though the mesogenic constituents are able to develop a long range orientation order [75,76].

8.4 Liquid Crystalline Polymer Nanocomposites

Change in the molecular structure of the liquid crystalline polymers can be achieved by forming nanocomposites with a range of nanoparticles. [77-84]. Quantum size effects, increased surface area, enhancement in the active surface, etc., are some of the properties achieved by introducing nanoparticles to the LCP matrices [85,86].

Grafting to polymerization was used for functionalizing CdTe, ZnO, SnO_2 and TiO_2 nanorods with polymers such as poly(methylmethacrylate), polystyrene and poly(diethylene glycol monomethyl ether) methacrylate deblock copolymers containing anchor groups [87]. Reversible addition-fragmentation chain transfer (RAFT) polymerization was used to synthesize the block copolymers. The surface coverage of the nanorods was observed to be dependent on the block lengths and ratios, as determined from the thermogravimetric analysis. The maximum surface coverage was noted for the short anchor unit blocks. Stable dispersions formed from the hairy rod like structures exhibited liquid crystalline behavior in different solvents and in polymeric and oligomeric matrices. In another study, the side chain liquid crystalline siloxanes doped with magnetic cobalt nanoparticles were studied for their structural and orientation properties [88]. At room temperature, the materials exhibited a combination of both ferromagnetic properties and orientational behavior. The nanorods were observed to self-assemble in bundles consisting of rows packed in a 2-dimensional hexagonal lattice (Figure 8.1).

Zorn and Zentel [89] reported the use of RAFT copolymerization to synthesize narrowly distributed block copolymers consisting of a hole conducting triarylamine block and an anchor block. Reactive ester approach was used to introduce the anchor block. The dopamine anchor groups containing copolymers were observed to bind to different semiconductors like ZnO, SnO_2 and TiO_2. Stable dispersions in particular solvents were achieved for the poly(triphenylamine) grafted inorganic nanorods. The nanorods demonstrated liquid crystalline behavior at higher concentration in various solvents and an oligotriphenylamine matrix.

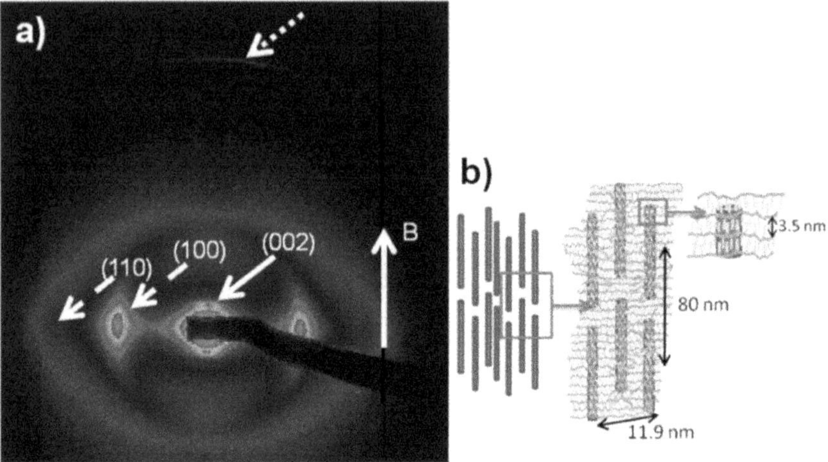

Figure 8.1 (a) Small angle X-ray scattering pattern of the Co nanorods hybrid material. The solid arrow points to the (001) reflection due to the organization of the Co nanorods in rows within the bundles whereas the dashed arrows point to the (100) and (110) reflections of the hexagonal 2-dimensional lattice and (b) schematic representation of the nanorod bundles. Reproduced from Reference 88 with permission from American Chemical Society.

Numerous other literature studies have also reported the development of new LCs and LCP composite materials [90-99]. Jang and Bae [93] reported the liquid crystalline epoxy/polyaniline (LCE/PANI) composite nanowires using an anodic aluminum oxide membrane. A temperature-gradient curing process was employed by the authors to develop the nanocomposites. Ezhov *et al.* [94] reported the nanocomposites of smectic C hydrogen-bonded polymers from the family of poly(4-(n-acryloyloxyalkoxy)benzoic acids with CdS nanorods (Figure 8.2). Uniaxial deformation of the composite films exhibited a long-range orientation with the formation of one-dimensional aggregates of rods. In another study, Horsch *et al.* [97] reported the self-assembly of nanorods functionalized by a polymer "tether." Meuer *et al.* [98] also functionalized TiO_2 nanorods with dopamine-functionalized diblock copolymers and observed the liquid crystalline behavior.

CNTs are attractive materials for reinforcing the host polymer matrices due to the properties like high aspect ratio, unique structural arrangement of atoms, outstanding mechanical, thermal and electrical performance, etc. [100-118]. The thermodynamic instability of thermotropic liquid crystalline polymers makes them immiscible with most thermoplastics at molecular level. The blend with poor mechanical properties are, thus, generated due to the poor interfacial adhesion between the polymers. The interfacial adhesion between the thermoplastic and liquid crystalline polymers can be improved by the using functionalized CNTs. This helps in improving the mechanical properties of the polymer blend, along with developing superior electrical and chemical properties [120].

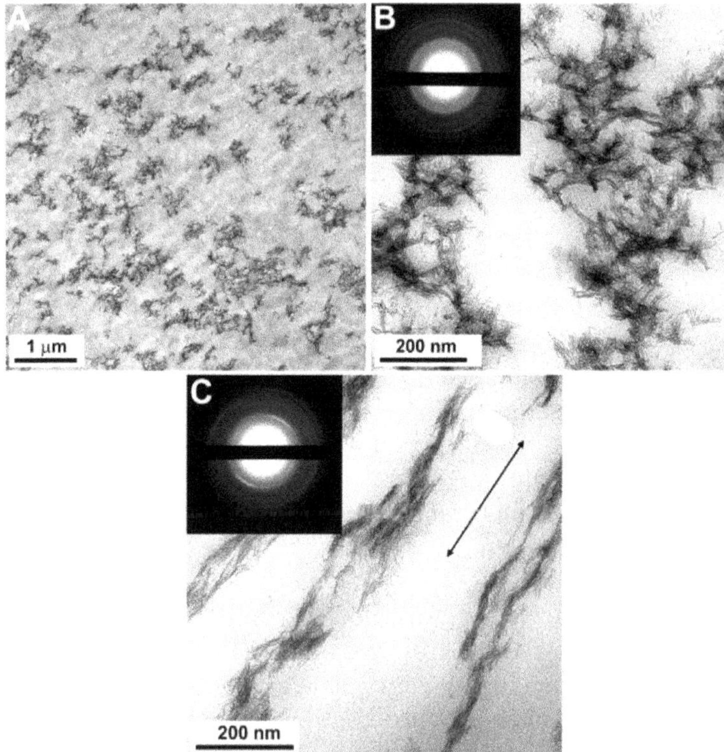

Figure 8.2 Transmission electron images of the composite containing 6.5 vol% of CdS nanorods before (A,B) and after the uniaxial deformation (C). Reproduced from Reference 94 with permission from American Chemical Society.

Mrozek and Taton [121] reported the LCP materials by seeding the growth of the LCP domains with the single-wall carbon nanotubes (SWNTs). The application of an electric field across the polyoxazoline melt incorporated with SWNTs seeds resulted in the orientation of the seeds (Figure 8.3). Subsequent cooling of the material resulted in the domains oriented in the direction of the applied field. Only a small amount of CNTs was required due to their action as nucleating agents.

Scheme 1

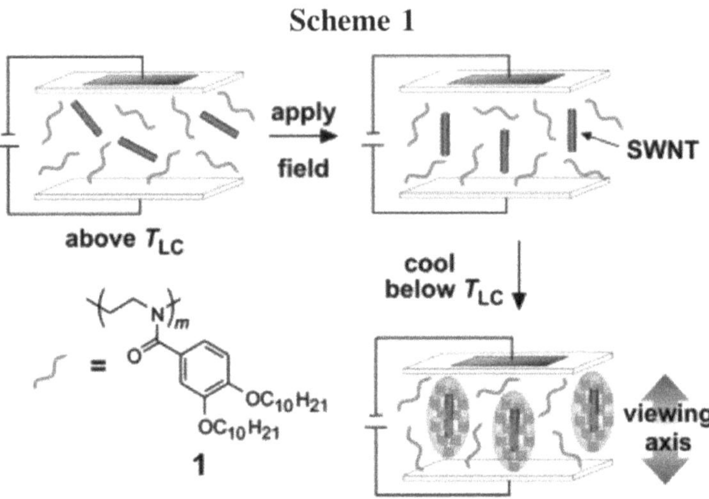

Figure 8.3 Schematic of the synthesis of the nanohybrid. Reproduced from Reference 121 with permission from American Chemical Society.

Display technologies have been the prime users of the liquid crystalline materials in the past decade. Bio-medical devices, light modulators, sensors and organic transistors are some of the other areas of commercial applications of these polymers [122-130]. In most of the LCP composites, surface modification of nanoparticles is required for generating interfacial compatibilization with the polymer phase. This paves the way of developing nanoparticles which are coated with LC ligands. The ligand exchange process is widely used for this purpose [130-136].

A number of studies have reported the *in-situ* synthesis of nanoparticles in the LC matrices for the purpose of developing nanocomposites. This involves the subsequent modifications of the precursors using techniques such as hydrolysis, oxidization and reduction. However, this approach faces challenges with respect to the compatibility of the precursor with the liquid crystal phase [136-151].

8.5 Conclusion

Many liquid crystalline polymer nanocomposite systems, exhibiting superior structural properties, have been developed in the recent past. Though many existing challenges currently hinder the full utilization of the potential of such materials for a variety of commercial applications, however, the ongoing research efforts in this direction are expected to further expand their widespread use in the near future.

References

1. *Liquid Crystalline Polymers: Synthesis, Properties and Applications*, Mittal, V. (ed.), Central West Publishing, Australia (2018).
2. Khoo, I.-C., and Wu, S.-T (1993) *Optics and Nonlinear Optics of Liquid Crystals*, World Scientific, Singapore.
3. *Recent Advances in Liquid Crystalline Polymers*, Chapoy, L. L. (ed.), Springer, Germany (1985).
4. Blumstein, A. (1985) *Polymeric Liquid Crystals*, Springer, Germany.
5. Griffin, A. C. (1984) *Liquid Crystals and Ordered Fluids*, 4th volume, Springer, Germany.
6. Xu, Q.-W., Man, H.-C. and Lau, W.-S. (1998) Study on the morphology and mechanical properties of binary and ternary blends of semiaromatic LCP/PP/PC. *Polymer-Plastics Technology and Engineering*, **37**, 253-259.
7. Lizuka, E. (2000) Effects of magnetic fields on polymer liquid crystals. *International Journal of Polymeric Materials*, **45**, 191-238.
8. Broer, D. J., Finkelmann, H., and Kondo, K. (1988) In-situ photopolymerization of an oriented liquid-crystalline acrylate. *Macromolecular Chemistry and Physics*, **189**, 185-194.
9. Broer, D. J., Mol, G. N., and Challa, G. (1989) In situ photopolymerization of an oriented liquid-crystalline acrylate, 2. *Macromolecular Chemistry and Physics*, **190**, 19-30.
10. Broer, D. J., Boven, J., Mol, G. N., and Challa, G. (1989) In-situ photopolymerization of oriented liquid-crystalline acrylates, 3. Oriented polymer networks from a mesogenic diacrylate. *Macromolecular Chemistry and Physics*, **190**, 2255-2268.
11. Hikmet, R. A. M., and Lub, J. (1995) Anisotropic networks with stable dipole orientation obtained by photopolymerization in the ferroelectric state. *Journal of Applied Physics*, **77**, 6234-6238.
12. He, L., Zhang, S., Jin, S., and Qi, Z. (1995) In-situ photopolymerization of an oriented chiral liquid crystal acrylate in an electric field. *Polymer Bulletin*, **34**, 7-12.
13. Hoyle, C. E., Mathias, L. J., Jariwala, C., and Sheng, D. (1996) Photopolymerization of a semifluorinated difunctional liquid crystalline monomer in a smectic phase. *Macromolecules*, **29**, 3182-3187.

14. Guymon, C. A., Hoggan, E. N., Clark, N. A., Rieker, T. P., Walba, D. M., and Bowman, C. N. (1997) Effects of monomer structure on their organization and polymerization in a smectic liquid crystal. *Science*, **275**, 57-59.

15. Broer, D. J., Mol, G. N., and Challa, G. (1991) Temperature effects on the kinetics of photoinitiated polymerization of dimethacrylates. *Polymer*, **32**, 690-695.

16. Hoyle, C. E., Kang, D., Jariwala, C., and Griffin, A. C. (1993) Efficient polymerization of a semi-fluorinated liquid crystalline methacrylate. *Polymer*, **34**, 3070-3075.

17. Hoyle, C. E., and Watanabe, T. (1994) Kinetics of polymerization of liquid-crystalline monomers: an exotherm and light scattering analysis. *Macromolecules*, **27**, 3790-3796.

18. Hoyle, C. E., Watanabe, T., and Whitehead, J. B. (1994) Anisotropic network formation by photopolymerization of liquid crystal monomers in a low magnetic field. *Macromolecules*, **27**, 6581-6588.

19. Sahlen, F., Trollsas, M., Hult, A., and Gedde, U. W. (1996) Synthesis and characterization of bifunctional liquid-crystalline monomers showing smectic C phase. photopolymerization and crosslinking. *Chemistry of Materials*, **8**(2), 382-388.

20. Hellermark, C., Gedde, U. W., and Hult, A. (1992) Synthesis and characterization of poly(11-(4'-cyano-trans-4-stilbenyloxy)undecanyl vinyl ether). *Polymer Bulletin*, **28**, 267-274.

21. Andersson, H., Gedde, U. W., and Hult, A. (1992) Preparation of ordered, crosslinked and thermally stable liquid crystalline poly(vinyl ether) films. *Polymer*, **33**, 4014-4018.

22. Broer, D. J., Lub, J., and Mol, G. N. (1993) Synthesis and photopolymerization of a liquid-crystalline diepoxide. *Macromolecules*, **26**, 1244-1247.

23. Jahromi, S., Lub, J., and Mol, G. N. (1994) Synthesis and photoinitiated polymerization of liquid crystalline diepoxides. *Polymer*, **35**, 622-629.

24. Andersson, H., Trollsas, M., Gedde, U. W., and Hult, A. (1995) Preparation of ordered and crosslinked films from liquid crystalline p-vinylphenoxy-based monomers. *Macromolecular Chemistry and Physics*, **196**, 3667-3676.

25. D'allest, J. F., Sixou, P., Blumstein, A., and Blumstein, R. B. (1988) Investigation of the nematic-isotropic biphase in thermotropic main chain polymers. Homogeneity of the pure nematic and isotropic phases. Part I: Microscopy and fractionation. *Molecular Crystals and Liquid Crystals Incorporating Nonlinear Optics*, **157**, 229-251.

26. Collyer A. A. (1996) Introduction to liquid crystal polymers. In: *Rheology and Processing of Liquid Crystal Polymers*, Acierno, D., Collyer, A. A. (eds), 2nd volume, Springer, Netherlands, doi: 10.1007/978-94-009-1511-4_1.

27. Thomas, E. L., Wood, B. A. (1985) Mesophase texture and defects in thermotropic liquid-crystalline polymers. *Faraday Discussions of the Chemical Society*, **79**, 229-239.

28. Vyas, R., Rida, A., Bhattacharya, S., and Tentzeris, M. M. (2007) Liquid Crystal Polymer (LCP): The Ultimate Solution for Low-Cost RF Flexible Electronics and Antennas. *Proceedings of the Polymers in Defense and Aerospace Applications*, France, p. 21.

29. Chung, T. S., Calundann, G. W., and East, A. J. (1989) Liquid-crystalline polymers and their applications. In: *Encyclopedia of Engineering Materials*, Chermisinoff, N. P. (ed.), Marcel Dekker, USA.

30. *High Performance Fibers*, Hearle, J. W. S. (ed.), Woodhead Publishing, USA (2001).

31. Carracher, Jr., C. E. (2017) *Polymer Chemistry*, 10th edition, CRC Press, USA.

32. *Liquid Crystals (Topics in Physical Chemistry)*, Stegemeyer, H. (ed.), Springer, Germany (1994).

33. *Liquid Crystals and Plastic Crystals*, Gray, G. W., and Winsor, P. A. (eds.), Ellis Horwood, UK (1974).

34. Robinson, C. (1956) Liquid-crystalline structures in solutions of a polypeptide. *Transactions of the Faraday Society*, **52**, 571-592.

35. Hijo, A. A. C. T., Maximo, G. JCosta, M. C., Cunha, R. L., Pereira, J. F. B., Kurnia, K. A., Batista, E. A. C., and Meirelles, A. J. A. (2017) Phase behavior and physical properties of new biobased ionic liquid crystals. *The Journal of Physical Chemistry B*, **121**(14), 3177-3189.

36. Ball, Z. T., Sivula, K., and Frechet, J. M. J. (2006) Well-defined fullerene-containing homopolymers and diblock copolymers with high fullerene content and their use for solution-phase and bulk organization. *Macromolecules*, **39**, 70-72.

37. Berg, S., Krone, V., and Ringsdorf, H. (1986) Structural variations of liquid-crystalline polymers: crossshaped and laterally linked mesogens in main chain and side group polymers. *Macromolecular Rapid Communications*, **7**, 381-388.

38. Chai, C. P., Zhu, X. Q., Wang, P., Ren, M. Q., Chen, X. F., Xu, Y. D., Fan, X. H., Ye, C., Chen, E. Q., and Zhou, Q. F. (2007) Synthesis and phase structures of mesogen-jacketed liquid crystalline polymers containing 1,3,4-oxadiazole based side chains. *Macromolecules*, **40**, 9361-9370.

39. Chandrasekhar, S., Sadashiva, B. K., and Suresh, K. A. (1977) Liquid crystals of disc-like molecules. *Pramana*, **9**(5), 471-480.

40. Chen, X., Tenneti, K. K., Li, C. Y., Bai, Y., Wan, X., Fan, X., Zhou, Q.-F., Rong, L., and Hsiao, B. S. (2007) Sidechain liquid crystalline poly(meth)acrylates with bent-Core mesogens. *Macromolecules*, **40**, 840-848.

41. Chen, X. F., Shen, Z., Wan, X. H., Fan, X. H., Chen, E. Q., Ma, Y., and Zhou, Q. F. (2010) Mesogen-jacketed liquid crystalline polymers.

Chemical Society Reviews, **39**, 3072-3101.

42. Godovsky, Y. K., Papkov, V. S., and Dusek, K. (1989) Thermotropic mesophases in element-organic polymers. *Advances in Polymer Science*, **88**, doi: 10.1007/BFb0017966.

43. Gopalan, P., Andruzzi, L., Li, X. F., and Ober, C. K. (2002) Fluorinated mesogen-jacketed liquid-crystalline polymers as surface-modifying agents: Design, synthesis and characterization. *Macromolecular Chemistry and Physics*, **203**, 1573-1583.

44. Mark. H. F. (1988) *Encyclopedia of Polymer Science and Engineering*, 9th volume, Wiley, USA.

45. Hessel, F., and Finkelmann, H. (1985) A new class of liquid crystal side chain polymers mesogenic groups laterally attached to the polymer backbone. *Polymer Bulletin*, **14**, 375-378.

46. Keith, C., Reddy, R. A., and Tschierske, C. (2005) The first example of a liquid crystalline side-chain polymer with bent-core mesogenic units: ferroelectric switching and spontaneous achiral symmetry breaking in an achiral polymer. *Chemical Communications*, **7**, 871-873.

47. Keith, C., Dantlgraber, G., Amaranatha Reddy, R., Baumeister, U., and Tschierske, C. (2007) Ferroelectric and antiferroelectric smectic and columnar liquid crystalline phases formed by silylated and non-silylated molecules with fluorinated bent cores. *Chemistry of Materials*, **19**(4), 694-710.

48. Chen, W.-H., Chuang, W.-T., Jeng, U.-S., Sheu, H.-S., and Lin, H.-C. (2011) New SmCG phases in a hydrogen-bonded bent-core liquid crystal featuring a branched siloxane terminal group. *Journal of American Chemical Society*, **133**(39), 15674-15685.

49. Westphal, E., Gallardo, H., Sebastian, N., Eremin, A., Prehm, M., Alaasar, M., and Tschierske, C. (2019) Liquid crystalline self-assembly of 2,5-diphenyl-1,3,4-oxadiazole based bent-core molecules and the influence of carbosilane end-groups. *Journal of Materials Chemistry C*, in print.

50. Jakli, A., Lavrentovich, O. D., and Selinger, J. V. (2018) Physics of liquid crystals of bent-shaped molecules. *Reviews of Modern Physics*, **90**, 045004.

51. Katranchev, B., and Petrov, M. (2016) Phase transitions in nanocomposites of hydrogen-bonded dimeric liquid crystals with mesogenic and non-mesogenic components. *Phase Transitions*, **89**(2), 115-132.

52. Pugh, C., Bae, J. Y., Dharia, J., Ge, J. J., and Cheng, S. Z. D. (1998) Induction of smectic layering in nematic liquid crystals using immiscible components. 2. laterally attached side-chain liquid-crystallin poly(norbornene)s and their low-molar-mass analogues with hydrocarbon/oligodimethylsiloxane substituents. *Macromolecules*, **31**, 5188-5200.

53. Saravanan, C., and Kannan, P. (2010) Fluorine-substituted azoben-

zene destabilizes polar form of optically switchable fulgimide unit in copolymer system. *Journal of Polymer Science, Part A: Polymer Chemistry*, **48**, 1565-1578.

54. Saravanan, C., Senthil, S., and Kannan, P. (2008) Click chemistry-assisted triazole-substituted azobenzene and fulgimide units in the pendant-based copoly(decyloxymethacrylate)s for dual-mode optical switches. *Journal of Polymer Science, Part A: Polymer Chemistry*, **46**, 7843-7860.

55. Vix, A., Stocker, W., Stamm, M., Wilbert, G., Zentel, R., and Rabe, J. P. (1998) Chain folding in liquid-crystalline main-chain polymers with a smectic phase. *Macromolecules*, **31** (26), 9154-9156.

56. Reck, B., and Rangsdorf, H. (1985) Combined liquid crystalline polymers, mesogens in the main chain and as side groups. *Macromolecular Rapid Communications*, **6**, 291-299.

57. Reddy, R. A., and Sadashiva, B. K. (2003) Influence of fluorine substituent on the mesomorphic properties of five-ring ester banana-shaped molecules. *Liquid Crystals*, **30**, 1031-1050.

58. Reddy, G. S. M., Narasimhaswamy, T., Jayaramudu, J., Sadiku, E. R., Raju, K. M., Ray, S. S. (2013) A new series of two-ring-based side chain liquid crystalline polymers, synthesis and mesophase characterization. *Australian Journal of Chemistry*, **66**, 667-675.

59. Tenneti, K. K., Chen, X., Li, C. Y., Shen, Z., Wan, F. X., Zhou, Q. F., Rong, L., and Hsiao, B. S. (2009) Influence of LC content on the phase structures of side-chain liquid crystalline block copolymers with bent-core mesogens. *Macromolecules*, **42**, 3510-3517.

60. Komori, T., and Shinkai, S. (1993) Novel columnar liquid crystals designed from cone-shaped calix[4] arenes. The rigid bowl is essential for the formation of the liquid crystal phase. *Chemistry Letters*, **22**, 1455-1458.

61. Kostromin, S. G., Shibaev, V. P., and Plate, N. A. (1987) Thermotropic liquid-crystalline polymers XXVI. Synthesis of comb-like polymers with oxygen containing spacers and a study of their phase transitions. *Liquid Crystals*, **2**, 195-200.

62. Lecommandoux, S., Klok, H. A., Sayar, M., and Stupp, S. I. (2003) Synthesis and self-organization of rod–dendron and dendron–rod–dendron molecules. *Journal of Polymer Science, Part A: Polymer Chemistry*, **41**, 3501-3518.

63. Lehn, J. M. (1988) Supramolecular chemistry, scope and perspectives molecules supramolecular and molecular devices. *Angewandte Chemie, International Edition*, **27**, 89-112.

64. Tamm, L. K., and McConnell, H. M. (1985) Supported phospholipid bilayers. *Biophysical Journal*, **47**(1), 105-113.

65. Maier, G. (2001) Low dielectric constant polymers for microelectronics. *Progress in Polymer Science*, **26**, 3-65.

66. Matsuo, Y., Muramatsu, A., Hamasaki, R., Mizoshita, N., Kato, T., and

Nakamura, E. (2004) Regioselective synthesis of 1,4-Di(organo)[60]fullerenes through DMF-assisted monoaddition of silylmethyl grignard reagents and subsequent alkylation reaction. *Journal of American Chemical Society*, **126**, 432-433.

67. Niori, T., Sekine, T., Watanabe, J., Furukawa, T., and Takezoe, H. (1996) Distinct ferroelectric smectic liquid crystals consisting of banana shaped achiral molecules. *Journal of Materials Chemistry*, **6**, 1231-1233.

68. Noel, C., and Navarad, P. (1991) Liquid crystal polymers. *Progress in polymer science*, **16**, 55-110.

69. Ortega, J., Folcia, C. L., Etxebarria, J., Gimeno, N., and Ros, M. B. (2003) Interpretation of unusual textures in the B2 phase of a liquid crystal composed of bent-core molecules. *Physical Review E*, **68**, 011707.

70. Shibaev, V. P. (2009) Liquid–crystalline polymers, past, present, and future. *Polymer Science Series A*, **51**, 1131.

71. Shibaev, V. P., and Plate, N. A. (1984) Thermotropic liquid-crystalline polymers with mesogenic side groups. *Advances in Polymer Science*, **60/61**, doi: 10.1007/3-540-12994-4_4.

72. Shibaev, V. P., and Plate, N. A. (1985) Synthesis and structure of liquid-crystalline side-chain polymers. *Pure and Applied Chemistry*, **57**, 1589-1602.

73. Spassky, N., Lacoudre, N., Le Borgne, A., Varion, J. P., Jun, C. L., Friedrich, C., and Noel, C. (1989) Liquid crystal polymers with terminally 1- phenyl 2-(4-cyanophenyl)-ethane substituted side chains. *Macromolecular Symposia*, **24**, 271-281.

74. Sawamura, M., Kawai, K., Matsuo, Y., Kanie, K., Kato, T., and Nakamura, E. (2002) Stacking of conical molecules with a fullerene apex into polar columns in crystals and liquid crystals. *Nature*, **419**, 702-705.

75. Siva Mohan Reddy G., Jayaramudu J., Ray S. S., Varaprasad K., and Rotimi Sadiku E. (2015) Side chain liquid crystalline polymers: Advances and applications. In: *Liquid Crystalline Polymers*, Thakur, V., and Kessler, M. (eds.), Springer, Germany, doi: 10.1007/978-3-319-20270-9_16.

76. Ahn, S.-k., Deshmukh, P., Gopinadhan, M., Osuji, C. O., and Kasi, R. M. (2011) Side-chain liquid crystalline polymer networks, exploiting nanoscale smectic polymorphism to design shape-memory polymers. *ACS Nano*, **5**(4), 3085-3095.

77. Percec, V., and Yourd, R. (1998) Liquid crystalline polyethers and copolyethers based on conformational isomerism. 3. The influence of thermal history on the phase transitions of the thermotropic polyethers and copolyethers based on 1-(4-hydroxyphenyl)-2-(2-methyl-4-hydroxyphenyl)ethane and flexible spacers containing an odd number of methylene units. *Macromolecules*, **22**, 3229–3242.

78. Percec, V., Rodrguez-Prarada, J. M., and Ericsson, C. (1987) Synthesis and characterization of liquid crystalline poly (p-vinylbenzyl ether)s. *Polymer Bulletin*, **17**, 347-352.

79. Percec, V., Heck, J., and Ungar, G. (1991) Liquid-crystalline polymers containing mesogenic units based on half-disk and rodlike moieties. 5. side-chain liquid-crystalline poly(methylsiloxanes) containing hemiphasmidic mesogens based on 4-[[3,4,5-tris(alkan-l-yloxy) benzoyl]oxy]-4'- [[p-(propan-1-yloxy)-benzoyl] oxy] biphenyl groups. *Macromolecules*, **24**, 4957-4962.

80. Percec, V., Heck, J., Lee, M., Ungar, G., and Alvarez-Castillo, A. (1992) Poly{2-vinyloxyethyl 3,4,5-tris [4-(n-dodecanyloxy)benzyloxy]benzoate}, a self-assembled supramolecular polymer similar to tobacco mosaic virus. *Journal of Materials Chemistry*, **2**, 1033-1039.

81. Percec, V., Schlueter, D., Ronda, J. C., Johansson, G., Ungar, G., and Zhou, J. P. (1996) Tubular architectures from polymers with tapered side groups. assembly of side groups via a rigid helical chain conformation and flexible helical chain conformation induced via assembly of side groups. *Macromolecules*, **29**, 1464-1472.

82. Percec, V., Ahn, C. H., Ungar, G., Yeardley, D. J. P., Moller, M., and Sheiko, S. S. (1998) Controlling polymer shape through the self-assembly of dendritic side-groups. *Nature*, **391**, 161-164.

83. Percec, V., Holerca, M. N., Uchida, S., Cho, W. D., Ungar, G., Lee, Y., and Yeardley, D. J. P. (2002) Exploring and expanding the three-dimensional structural diversity of supramolecular dendrimers with the aid of libraries of alkali metals of their AB3 minidendritic carboxylates. *Chemistry, A European Journal*, **8**, 1106-1117.

84. Percec, V., Rudick, J. G., Peterca, M., Wagner, M., Obata, M., Mitchell, C. M., Cho, W. D., Balagurusamy, V. S. K., and Heiney, P. A. (2005) Thermoreversible cis-cisoidal to cis-transoidal isomerization ofhelical dendronized polyphenylacetylenes. *Journal of American Chemical Society*, **127**, 15257-15264.

85. Piunova, V. A., Miyake, G. M., Daeffler, C. S., Weitekamp, R. A., and Grubbs, R. H. (2013) Highly ordered dielectric mirrors via the self-assembly of dendronized block copolymers. *Journal of American Chemical Society*, **135**, 15609-15616.

86. Pugh, C., and Schrock. R. R. (1992) Synthesis of side-chain liquid crystal polymers by living ring-opening metathesis polymerization. 3. influence of molecular weight, interconnecting unit, and substituent on the mesomorphic behavior of polymers with laterally attached mesogens. *Macromolecules*, **25**, 6593-6604.

87. Zorn, M., Meuer, S., Tahir, M. N., Khalavka, Y., Sönnichsen, C., Tremel, W., and Zentel, R. (2008) Liquid crystalline phases from polymer functionalised semiconducting nanorods. *Journal of Materials Chemistry*, **18**, 3050-3058.

88. Riou, O., Lonetti, B., Davidson, P., Tan, R. P., Cormary, B., Mingotaud, A.-F., Di Cola, E., Respaud, M., Chaudret, B., Soulantica, K., and Mauzac, M. (2014) Liquid crystalline polymer–Co nanorod hybrids, structural analysis and response to a magnetic field. *The Journal of Physical Chemistry B*, **118** (11), 3218-3225.

89. Zorn, M., and Zentel, R. (2008) Liquid crystalline orientation of semiconducting nanorods in a semiconducting matrix. *Macromolecular Rapid Communications*, **29**, 922-927.

90. Meuer, S., Oberle, P., Theato, P., Tremel, W. and Zentel, R. (2007) Liquid crystalline phases from polymer-functionalized TiO$_2$ nanorods. *Advanced Materials*, **19**, 2073-2078.

91. Li, L.-S., and Alivisatos, A. P. (2003) Semiconductor nanorod liquid crystals and their assembly on a substrate. *Advanced Materials*, **15**, 408-411.

92. Yoshino, T., Kondo, M., Mamiya, J.-I., Kinoshita, M., Yu, Y. and Ikeda, T. (2010) Three-dimensional photomobility of crosslinked azobenzene liquid-crystalline polymer fibers. *Advanced Materials*, **22**, 1361-1363.

93. Jang, J., and Bae, J. (2005) formation of polyaniline nanorod/liquid crystalline epoxy composite nanowires using a temperature-gradient method. *Advanced Function Materials*, **15**, 1877-1882.

94. Ezhov, A. A., Shandryuk, G. A., Bondarenko, G. N., Merekalov, A. S. Abramchuk, S. S., Shatalova, A. M., Manna, P., Zubarev, E. R., and Talroze, R. V. (2011) Liquid-crystalline polymer composites with CdS nanorods, structure and optical properties. *Langmuir*, **27**(21), 13353-13360.

95. Nikhil, R., Gearheart, J. L. A., Obare, S. O., Johnson, C. J., Edler, K. J., Mann S., and Murphy, C. J. (2002) Liquid crystalline assemblies of ordered gold nanorods. *Journal of Materials Chemistry*, **12**, 2909-2912.

96. Dessombz, A., Chiche, D., Davidson, P., Panine, P., Chanéac, C., and, and Jolivet, J.-P. (2007) design of liquid-crystalline aqueous suspensions of rutile nanorods, evidence of anisotropic photocatalytic properties. *Journal of American Chemical Society*, **129**(18), 5904-5909.

97. Horsch, M. A., Zhang, Z., and Glotzer, S. C. (2005) Self-assembly of polymer-tethered nanorods. *Physical Review Letters*, **95**(5), 056105.

98. Meuer, S. Fischer, K. Mey, I. Janshoff, A. Schmidt, M. and Zentel, R. (2008) Liquid crystals from polymer-functionalized TiO$_2$ nanorod mesogens. *Macromolecules*, **41**(21), 7946-7952.

99. Guo, L., Cheng, J. X., Li, X.-Y., Yan, Y. J., Yang, S. H., Yang, C. L., Wang, J. N., and Ge, W. K. (2001) Synthesis and optical properties of crystalline polymer-capped ZnO nanorods, *Materials Science and Engineering C*, **16**(1-2), 123-127.

100. *Liquid Crystal Polymers*, Plate, N. A. (ed.), Plenum Press, USA (1992).
101. *Advances in Liquid Crystals*, Brown, G. H. (ed.), Academic Press, USA (1975).
102. Collings, P. J. (1990) *Liquid Crystals, Nature's Delicate Phase of Matter*, Princeton University Press, USA.
103. Tjong, S. C. (2003) Structure, morphology, mechanical and thermal characteristics of the in situ composites based on liquid crystalline polymers and thermoplastics. *Materials Science and Engineering R: Reports*, **41**, 1-60.
104. Donald, A. M., and Windle, A. H. (1992) *Liquid Crystalline Polymers*, Cambridge University Press, UK.
105. *Surface Modification of Nanotube Fillers*, Mittal, V. (ed.), Wiley VCH, Germany (2011).
106. *Polymer-Graphene Nanocomposites*, Mittal, V. (ed.), Royal Society of Chemistry, UK, 2012.
107. Iijima, S. (1991) Helical microtubules of graphitic carbon. *Nature*, **354**, 56-58.
108. Demus, D., Gray, G. W., Speiss, H. W., Goodby, J. W., and Vill, V. (1998) *Handbook of Liquid Crystals*, Wiley-VCH, Germany.
109. Park, S. K., Kim, S. H., and Hwang, J. T. (2008) Carboxylated multiwall carbon nanotube-reinforced thermotropic liquid crystalline polymer nanocomposites. *Journal of Applied Polymer Science*, **109**, 388-396.
110. Kim, J. Y., Kim, D. K., and Kim, S. H. (2009) Effect of modified carbon nanotube on physical properties of thermotropic liquid crystal polyester composites. *European Polymer Journal*, **45**, 316-324.
111. Kaito, A., Kyotani, M., and Nakayama, K. (1990) Effects of annealing on the structure formation in a thermotropic liquid crystalline copolyester. *Macromolecules*, **23**, 1035-1040.
112. Moniruzzaman, M., and Winey, K. I. (2006) Polymer nanocomposites containing carbon nanotubes. *Macromolecules*, **39**, 5194-5205.
113. Xie, X. L., Mai, Y. W., and Zhou, X. P. (2005) Dispersion and alignment of carbon nanotubes in polymer matrix: A review. *Materials science and Engineering R: Reports*, **49**, 89-112.
114. Coleman, J. N., Khan, U., Blau, W. J., and Gun'ko, Y. K. (2006) Small but strong, A review of the mechanical properties of carbon nanotube-polymer composites. *Carbon*, **44**, 1624-1652.
115. Coleman, J. N., Khan, U., and Gun'ko, Y. K. (2006) Mechanical reinforcement of polymers using carbon nanotubes. *Advanced Materials*, **18**, 689-706.
116. Kiss, G. (1987) In situ composites, Blends of isotropic polymers and thermotropic liquid crystalline polymers. *Polymer Engineering and Science*, **27**, 410-423.
117. Lin, Y. G., and Winter, H. H. (1991) High-temperature recrystalliza-

tion and rheology of a thermotropic liquid crystalline polymer. *Macromolecules*, **24**, 2877-2882.

118. Serpe, G., and Economy, J. (1992) Ordering processes in the 2,6-hydroxynaphthoic acid (HNA) rich copolyesters of p-hydroxybenzoic acid and HNA. *Macromolecular Symposia*, **53**, 65-75.

119. Cheng, H. K. F., Sahoo, N. G., Li, L., Chan, S. H., and Zhao, J. (2010) Molecular interactions in PA6, LCP and their blend incorporated with functionalized carbon nanotubes. *Key Engineering Materials*, **447-448**, 634-638.

120. Cheng, H. K. F., Basu, T., Sahoo, N. G., Li, L., and Chan, S. H. (2012) Current advances in the carbon nanotube/thermotropic main-chain liquid crystalline polymer nanocomposites and their blends. *Polymers*, **4**(2), 889-912.

121. Mrozek, R. A., and Taton, T. A. (2005) Alignment of liquid-crystalline polymers by field-oriented, carbon nanotube directors. *Chemistry of Materials*, **17**(13), 3384-3388.

122. Dasgupta, D., Shishmanova, I. K., Ruiz-Carretero, A., Lu, K., Verhoeven, M., van Kuringen, H.P.C, Portale, G., Leclere, P., Bastiaansen, C. W. M., Broer, D. J., and Schenning, A. P. H. J. (2013) Patterned silver nanoparticles embedded in a nanoporous smectic liquid crystalline polymer network. *Journal of American Chemical Society*, **135**, 10922-10925.

123. Dobbs, W., Suisse, J. M., Douce, L., and Welter, R. (2006) Electrodeposition of silver particles and gold nanoparticles from ionic liquid-crystal precursors. *Angewandte Chemie International Edition*, **45**, 4179-4182.

124. Domenici, V., Zupancic, B., Laguta, V. V., Belous, A. G., V'Yunov, O. I., Remskar, M., Zalar, B. (2010) $PbTiO_3$ nanoparticles embedded in a liquid crystalline elastomer matrix, structural and ordering properties. *Journal of Physical Chemistry C*, **114**, 10782-10789.

125. Domenici, V., Conradi, M., Remskar, M., Virsek, M., Zupancic, B., Mrzel, A., Chambers, M., and Zalar, B. (2011) New composite films based on MoO_3-x nanowires aligned in a liquid single crystal elastomer matrix. *Journal of Materials Science*, **46**, 3639-3645.

126. Saliba, S., Coppel, Y., Achard, M. F., Mingotaud, C., Marty, J. D., and Kahn, M. L. (2011) Thermotropic liquid crystals as templates for anisotropic growth of nanoparticles. *Angewandte Chemie, International Edition*, **50**, 12032-12035.

127. Saliba, S., Coppel, Y., Mingotaud, C., Marty, J. D., and Kahn, M. L. (2012) ZnO/liquid crystalline nanohybrids, from solution properties to the control of the anisotropic growth of nanoparticles. *Chemistry, A European Journal*, **18**, 8084-8091.

128. Saliba, S., Mingotaud, C., Kahn, M. L., and Marty, J. D. (2013) Liquid crystalline thermotropic and lyotropic nanohybrids. *Nanoscale*, **5**, 6641-6661.

129. Shandryuk, G. A., Matukhina, E. V., Vasil'ev, R. B., Rebrov, A., Bondarenko, G. N., Merekalov, A. S., Gas'kov, A. M., and Talroze, R. V. (2008) Effect of H-bonded liquid crystal polymers on CdSe quantum dot alignment within nanocomposite. *Macromolecules*, **41**, 2178-2185.

130. Taubert, A. (2004) CuCl nanoplatelets from an ionic liquid-crystal precursor. *Angewandte Chemie, International Edition*, **43**, 5380-5382.

131. Vasilets, V. N., Savenkov, G. N., Merekalov, A. S., Shandryuk, G. A., Shatalova, A. M., and Tal'roze, R. V. (2011) Immobilization of quantum dots of cadmium selenide on the matrix of a graft liquid-crystalline polymer. *Polymer Science Series A*, **53**, 521-526.

132. Zadoina, L., Lonetti, B., Soulantica, K., Mingotaud, A. F., Respaud, M., Chaudret, B., and Mauzac, M. (2009) Liquid crystalline magnetic materials. *Journal of Materials Chemistry*, **19**, 8075-8078.

133. Zadoina, L., Soulantica, K., Ferrere, S., Lonetti, B., Respau, M., Mingotaud, A. F., Falqui, A., Genovese, A., Chaudret, B., and Mauzac, M. (2011) In situ synthesis of cobalt nanoparticles in functionalized liquid crystalline polymers. *Journal of Materials Chemistry*, **21**, 6988-6994.

134. Barmatov, E. B., Pebalk, D. A., and Barmatova, M. V. (2004) Influence of silver nanoparticles on the phase behavior of side-chain liquid crystalline polymers. *Langmuir*, **20**, 10868-10871.

135. Brochard, F., and de Gennes, P. G. (1970) Theory of magnetic suspensions in liquid crystals. *Journal de Physique*, **31**(7), 691-708.

136. Chambers, M., Zalar, B., Remskar, M., Zumer, S., and Finkelmann, H. (2006) Actuation of liquid crystal elastomers reprocessed with carbon nanoparticles. *Applied Physics Letters*, **89**, 243116.

137. Domracheva, N. E., Pyataev, A. V., Manapov, R. A., and Gruzdev, M. S. (2011) Magnetic resonance and M€ossbauer studies of superparamagnetic γ-Fe$_2$O$_3$ nanoparticles encapsulated into liquidcrystalline poly(propylene imine) dendrimers. *ChemPhysChem*, **12**, 3009-3019.

138. Garcia-Marquez, A., Demortiere, A., Heinrich, B., Guillon, D., Begin-Colin, S., and Donnio, B. (2011) Iron oxide nanoparticle-containing main-chain liquid crystalline elastomer, towards soft magnetoactive networks. *Journal of Materials Chemistry*, **21**, 8994-8996.

139. Gascon, I., Marty, J. D., Gharsa, T., and Mingotaud, C, (2005) Formation of gold nanoparticles in a sidechain liquid crystalline network, influence of the structure and macroscopic order of the material. *Chemistry of Materials*, **17**, 5228-5230.

140. Haberl, J. M., Sanchez-Ferrer, A., Mihut, A. M., Dietsch, H., Hirt, A. M., and Mezzenga, R. (2013) Liquid crystalline elastomer-nanoparticle hybrids with reversible switch of magnetic memory. *Advanced Materials*, **25**, 1787-1791.

141. Haberl, J. M., Sanchez-Ferrer, A., Mihut, A. M., Dietsch, H., Hirt, A. M., and Mezzenga, R. (2013) Straininduced macroscopic magnetic anisotropy from smectic liquid-crystalline elastomermaghemite nanoparticle hybrid nanocomposites. *Nanoscale*, **5**, 5539-5548.

142. Hegman, T., Qi, H., and Marx, V. M. (2007) Nanoparticles and liquid crystals, synthesis, self-assembly, defect formation and potential applications. *Journal of Inorganic and Organometallic Polymers and Materials*, **17**, 483-508.

143. Kaiser, A., Winkler, M., Krause, S., Finkelmann, H., and Schmidt, A. M. (2009) Magnetoactive liquid crystal elastomer nanocomposites. *Journal of Materials Chemistry*, **19**, 538-543.

144. Lee, J., and W., Jin, J. (2003) Formation of gold nanoparticles within a liquid crystalline polymeric matrix. *Journal of Nanoscience and Nanotechnology*, **3**, 219-221.

145. Li, F, Chen, W, and Chen, Y. W. (2012) Mesogen induced self-assembly for hybrid bulk heterojunction solar cells based on a liquid crystal D-A copolymer and ZnO nanocrystals. *Journal of Material Chemistry*, **22**, 6259-6266.

146. Mallia, V. A., Vemula, P. K., John, G., Kumar, A., and Ajayan, P. M. (2007) In situ synthesis and assembly of gold nanoparticles embedded in glass-forming liquid crystals. *Angewandte Chemie, International Edition*, **46**, 3269-3274.

147. Marshall, J. E., Ji, Y., Torras, N., Zinoviev, K., and Terentjev, E. M. (2012) Carbon-nanotube sensitized nematic elastomer composites for IR-visible photo-actuation. *Soft Matter*, **8**, 1570-1574.

148. Mertelj, A., Lisjak, D., Drofenik, M., and Copic, M. (2013) Ferromagnetism in suspensions of magnetic platelets in liquid crystal. *Nature*, **504**, 237-241.

149. Montazami, R., Spillmann, C. M., Naciri, J., and Ratna, B. R. (2012) Enhanced thermomechanical properties of a nematic liquid crystal elastomer doped with gold nanoparticles. *Sensors and Actuators A: Physical*, **178**, 175-178.

150. Nguyen, H. H., Valverde, S. C., Lavedan, P., Goudoune`che, D., Mingotaud, A. F., Lauth-de Viguerie, N., and Marty, J. D. (2014) Mesomorphic ionic hyperbranched polymers, effect of structural parameters on liquid-crystalline properties and on the formation of gold nanohybrids. *Nanoscale*, **6**, 3599-3610.

151. Riou, O., Zadoina, L., Lonetti, B., Soulantica, K., Mingotaud, A.-F., Respaud, M., Chaudret, B., and Mauzac, M. (2012) *In situ* and *ex situ* syntheses of magnetic liquid crystalline materials: A comparison. *Polymers,* **4**(1), 448-462.

9

Polymer Nanocomposite Inks and Pigments

9.1 Introduction

Significant research attention has been drawn towards printed electronics in the recent years owing to usefulness in flexible, large size and cost-effective electronics [1-7]. In this respect, a considerable progress has been made in printable semiconductor and electrode materials [7]. However, the dielectric layers, one of the most important constituents of thin-film transistors (TFTs), are not often printed due to the difficulty of printing a high-quality layer and the lack of printable dielectric materials [8]. Usually in the thin-film transistors, the thickness of dielectric films must be in the range of hundreds of nanometers to a few micrometers so as to avoid any leakage current. Additionally, in order to bring down the gate voltage, it is required that the printed dielectric layer has a large capacitance per unit area.

Various dielectric materials, based on solution processing, have been fabricated as gate dielectrics. These generally comprise of inorganic oxides [9], polymers [8], solid-state electrolytes [11] and iongels [10]. Solid state electrolytes and ion gels [6,7,12,13] have high dielectric constant and show enhanced conductivity at high frequency which restricts the transistor's switching speed. ZrO_2 [9] shows high frequency stability and permittivity, although the prevention of printed ceramic films from cracks and pinholes is difficult, leading to short circuits in appliances. However, polymer dielectric materials with low permittivity are relatively easier to print as compared to other materials. Composites integrating printable polymer and high-permittivity inorganic nanoparticles can result in novel dielectric materials having good printability, higher dielectric constant as well as high frequency stability. In such composite materials, the dielectric constant mostly elevates with the addition of high-permittivity inorganic nanoparticles. Nevertheless, with an increase in the concentration of nanoparticles, the leakage current enhances as well due to the electromigration of the metallic atoms via gaps from the

Swati Singh and Vikas Mittal, Khalifa University of Science and Technology, Abu Dhabi, UAE

electrode material. In this case, two-dimensional (2D) dielectric materials are more suitable as compared to the spherical nanoparticles owing to their lamellar geometry [14-16]. Overall, various high performance polymers have been employed to develop nanocomposite inks and pigments for printing [17-38].

9.2 Printing Techniques

In three-dimensional printing, powdered materials are provided on a platform, and a liquid binder based on water is spread using inkjet printing so as to aggregate the individual particles and shape a two-dimensional sequence [39]. Another layer of powder is then introduced on the surface of the printed layers using a roller. In order to develop a strong bonding between the powder particles, post-heating is used [39]. A large variety of materials such as polymer composites, ceramics and metals can be effectively printed using this method. Nevertheless, low printing resolution, high cost and non-optimal surface finish are concerns for the three-dimensional printing process [40,41].

Laminated object printing is an amalgam of additive and subtractive manufacturing [42,43]. On the other hand, the mechanism of stereolithography is based on a photo-instigated polymerization [44-47]. Other printing techniques include paste extrusion printing, selective laser sintering, polyjet, laser engineered net shaping, etc. [48-70].

9.3 Polymer Nanocomposites based Inks

Peak *et al.* [71] investigated the rheological modification of poly(ethyleneglycol) (PEG) precursor solutions via incorporation of laponite clay nanoparticles. PEG/laponite formed internal "house-of cards" structure, influencing fluid flow and ability to print (Figure 9.1). Laponite addition to PEG reduced the recovery time of solutions from infinite (Newtonian fluid) to seconds, which is more appropriate for bioprinting applications.

Titanium dioxide (TiO$_2$) nanoparticles are commonly employed in applications such as photocatalysts, batteries, electrochromic and photochromic devices, photovoltaic cells, gas sensors and thin-film transistors, due to the strong oxidizing power of the photogenerated holes, non-toxicity, high refractive index, chemical inertness and cost-effectiveness. N-doped TiO$_2$ nanoparticles were dissolved by

Loffredo *et al.* [72] in solution with the help of a polycation polymer (polyethyleneimine (PEI)) to prepare inks for ink jet printing applications. At different PEI concentrations, various TiO_2/PEI/EtOH suspensions were prepared, and the impact of the dispersant content on the ink printability was thoroughly investigated. The suspensions were observed to exhibit good chemico-physical behavior for their use as inks.

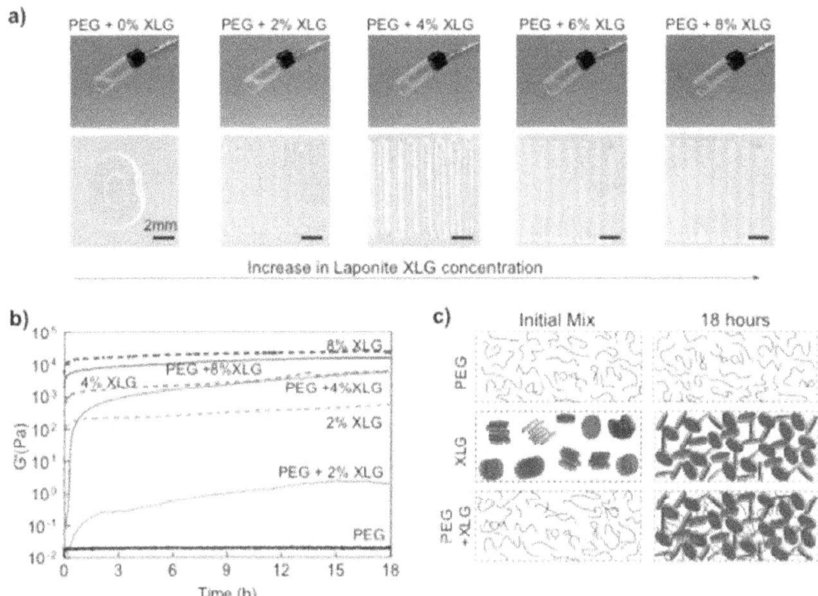

Figure 9.1 Synthesis of PEG-laponite colloidal solution: (a) viscosity increases at rest with increasing laponite concentration, (b) time sweep curves and (c) schematic of internal structure formation of PEG, laponite and PEG/laponite solutions. Reproduced from Reference 71 with permission from American Chemical Society.

In order to obtain an environment friendly and printable ink, a nanocomposite based on graphene and acrylic polymer was demonstrated [73]. Thin printing tests of the nanocomposite demonstrated a diminishing of resistivity by two orders of magnitude. Directly written TiO_2 films were fabricated by Arango *et al.* [74] on flexible substrates from aqueous systems. Mild temperature and UV irradiation conditions were used to transform the amorphous/crystalline formulations to semi-crystalline/crystalline films for diverse applications on flexible substrates. It was observed that the viscosity and printing properties of the inks could be tailored through solvent and polymer addition and titanium (IV) bis(ammonium lactato) dihydroxide (TALH):TiO$_2$ ratio (Figure 9.2).

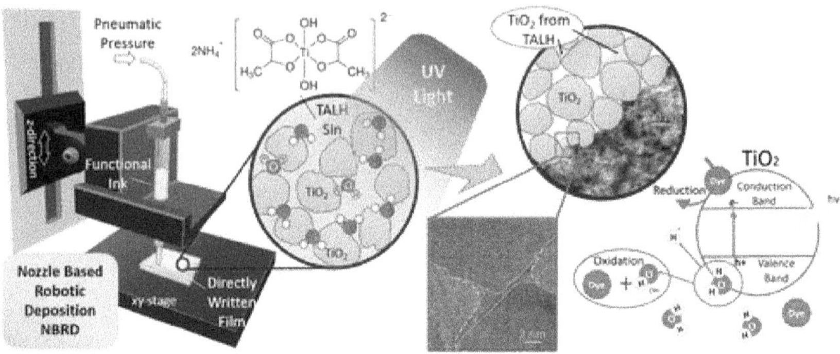

Figure 9.2 Schematic representation of the synthesis of printed films. Reproduced from Reference 74 with permission from American Chemical Society.

Natural composite materials were also developed over an expansive length scale by the immediate stage trans-development of all-fluid ink to composite material [75]. The variations in the degree of ink bead mixture amid printing were observed to create differing material morphologies.

9.4 Polymer Nanocomposites based Pigments

Natural shades have been widely utilized for paints and printing applications owing to their photosensitivity, shading quality, etc. Natural shades are hard to be wetted and dispersed in water or polymeric films [76,77]. In past decades, advancements in this area have been achieved through progression in heterogeneous polymerization and nanocomposites technology through emulsion polymerization [78,79], miniemulsion polymerization [80,81], suspension polymerization [82,83] or micro-suspension polymerization [84,85].

Qi *et al.* [86] reported improvement in dispersion and color of organic pigments in polymeric films via micro-encapsulation. Zeng *et al.* [87] also reviewed the applications of polymer nanocomposites in coatings and pigments. Films of PU/PVC containing multi-walled and single-walled carbon nanotubes (MWCNTs and SWCNTs) were prepared by Abdelrazek *et al.* [88]. In another study, nanocomposite films consisting of polyvinylidene fluoride (PVDF)/polyvinyl chloride (PVC) blend doped with 0.005 and 0.010 wt% of graphene oxide nanoparticles (GO) were reported [89].

Kiatkamjornwong and Pomsanam [83] reported styrenic-based polymerized toner and its composite for electrophotographic printing. The analysis of print quality exhibited high background fog, low maximum density and small extent of image raggedness. Nanoblue shade poly(styrene-co-n-butyl acrylate-co-methacrylic acid) composite particles were also reported by fast suspension polymerization by Widiyandari *et al.* [84]. In another study, simple and ecofriendly method was used to prepare titanium dioxide/polymer composites using compound oxidative polymerization [90]. Ghannam *et al.* [91] also reported colored diblock copolymer-mica composite pigments. The pigments were prepared from the mica surface by carrying out the block copolymerization of butyl acrylate and a mixture of styrene and a dye.

9.5 Conclusions

In this study, the development of polymer nanocomposite inks and pigments has been briefly reviewed. Owing to the much improved properties as compared to the conventional materials, these nanomaterials exhibit high potential of wide ranging applications in near future.

References

1. Wang, C., Qian, L., Xu, W., Nie, S., Gu, W., and Zhang, J. (2013) High performance thin film transistors based on regioregular poly(3-dodecylthiophene)-sorted large diameter semiconducting single-walled carbon nanotubes. *Nanoscale*, **5**, 4156-4161.
2. Han, S. Y., Lee, D. H., Herman, G. S., and Chang, C. H. (2009) Inkjet-printed high mobility transparent-oxide semiconductors. *Journal of Display Technology*, **5**, 520-524.
3. Yan, H., Chen, Z., Zheng, Y., Newman, C., Quinn, J. R., and Dötz, F. (2009) A high-mobility electron-transporting polymer for printed transistors. *Nature*, **457**, 679-686.
4. Woo, K., Bae, C., Jeong, Y., Kim, D., and Moon, J. (2010) Inkjet-printed Cu source/drain electrodes for solution-deposited thin film transistors. *Journal of Materials Chemistry*, **20**, 3877-3882.
5. Chung, S., Jang, M., Ji, S. B., Im, H., Seong, N., Ha, J., Kwon, S.-K., Kim, Y.-H., Yang, H., and Hong, Y. (2013) Flexible high-performance all-

inkjet-printed inverters: Organo-compatible and stable interface engineering. *Advanced Materials*, **25**, 4773-4777.

6. Zhao, J., Gao, Y., Gu, W., Wang, C., Lin, J., and Chen, Z. (2012) Fabrication and electrical properties of all-printed carbon nanotube thin film transistors on flexible substrates. *Journal of Materials Chemistry*, **22**, 20747-20753.

7. Zhao, J., Gao, Y., Lin, J., Chen, Z., and Cui, Z. (2012) Printed thin-film transistors with functionalized single-walled carbon nanotube inks. *Journal of Materials Chemistry*, **22**, 2051-2056.

8. Ortiz, R., Facchetti, A., and Marks, T. J. (2010) High-k organic, inorganic, and hybrid dielectrics for low-voltage organic field-effect transistors. *Chemical Reviews*, **110**, 205-239.

9. Park, J. H., Yoo, Y. B., Lee, K. H., Jang, W., Oh, J., and Chae, S. (2012) Low-temperature, high- performance solution-processed thin-film transistors with peroxo-zirconium oxide dielectric. *ACS Applied Materials & Interfaces*, **5**, 410-417.

10. Cho, J. H., Lee, J., He, Y., Kim, B. S., Lodge, T. P., and Frisbie, C. D. (2008) High-capacitance ion gel gate dielectrics with faster polarization response times for organic thin film transistors. *Advanced Materials*, **20**, 686-690.

11. Braga, D., Ha, M., Xie, W., and Frisbie, C. D. (2010) Ultralow contact resistance in electrolytegated organic thin film transistors. *Applied Physics Letters*, **97**, 193311.

12. Lee, J., Kaake, L. G., Cho, J. H., Zhu, X. Y., Lodge, T. P., and Frisbie, C. D. (2009) Ion gel-gated polymer thin-film transistors: operating mechanism and characterization of gate dielectric capacitance, switching speed, and stability. *Journal of Physical Chemistry C*, **113**, 8972-8981.

13. Hong, K., Kim, S. H., Lee, K. H., and Frisbie, C. D. (2013) Printed sub-2V ZnO electrolyte gated transistors and inverters on plastic. *Advanced Materials*, **25**, 3413-3418.

14. Jang, W. S., Rawson, I., and Grunlan, J. C. (2008) Layer-by-layer assembly of thin film oxygen barrier. *Thin Solid Films*, **516**, 4819-4825.

15. Wu, X., Fei, F., Chen, Z., Su, W., and Cui, Z. (2014) A new nanocomposite dielectric ink and its application in printed thin-film transistors. *Composites Science and Technology*, **94**, 117-122.

16. *Polymer Nanocomposites, Printable and Flexible Technology for Electronic Packaging*. Online: https://smtnet.com/library/files/upload/Printable-Nanocomposites.pdf [accessed 26th June 2019].

17. Yang, X., Loos, J., Veenstra, S. C., Verhees, W. J. H., Wienk, M. M., Kroon, J. M., Michels, M. A. J., and Janssen, R. A. J. (2005) Nanoscale morphology of high-performance polymer solar cells. *Nano Letters*, **5**, 580-582.

18. *High Performance Polymers and Engineering Plastics*, Mittal, V. (ed.),

John Wiley & Sons, USA (2011).

19. Hiemenz P. C., and Lodge, T. P. (2007) *Polymer Chemistry*, CRC Press, USA.

20. Cogswell, F. N. (2013) *Thermoplastic Aromatic Polymer Composites*, Elsevier, USA.

21. Molazemhosseini, A., Tourani, H., Naimi-Jamal, M. R., and Khavandi, A. (2013) Nanoindentation and nanoscratching responses of PEEK based hybrid composites reinforced with short carbon fibers and nano-silica. *Polymer Testing*, **32**, 525-534.

22. Díez-Pascual, A. M., Naffakh, M., Gonzalez-Domínguez, J. M., Anson, A., Martínez-Rubi, Y., Martinez, M. T., Simard, B., Gomez, M. A. (2010) High performance PEEK/carbon nanotube composites compatibilized with polysulfones-II. Mechanical and electrical properties. *Carbon*, **48**, 3500-3511.

23. LiuJie, X., Davim, J. P., Cardoso, R. (2007) Prediction on tribological behaviour of composite PEEK-CF30 using artificial neural networks. *Journal of Materials Processing Technology*, **189**, 374-378.

24. Nisa, V., Rajesh, S., Murali, K., Priyadarsini, V., Potty, S., and Ratheesh, R. (2008) Preparation, characterization and dielectric properties of temperature stable $SrTiO_3$/PEEK composites for microwave substrate applications. *Composites Science and Technology*, **68**, 106-112.

25. Iqbal, T., Briscoe, B., and Luckham, P. (2011) Scratch deformations of poly (etheretherketone). *Wear*, **271**, 1181-1193.

26. Iqbal, T., Briscoe, B. J., Yasin, S., and Luckham, P. F. (2013) Nanoindentation response of poly(ether ether ketone) surfaces da semicrystalline bimodal behavior. *Journal of Applied Polymer Science*, **130**, 4401-4409.

27. Powles, R., McKenzie, D., Meure, S., Swain, M., and James, N. (2007) Nanoindentation response of PEEK modified by mesh-assisted plasma immersion ion implantation. *Surface and Coatings Technology*, **201**, 7961-7969.

28. Díez-Pascual, A. M., Gomez-Fatou, M. A., Ania, F., and Flores, A. (2015) Nanoindentation in polymer nanocomposites. *Progress in Materials Science*, **67**, 1-94.

29. Godara, A., Raabe, D., and Green, S. (2007) The influence of sterilization processes on the micromechanical properties of carbon fiber-reinforced PEEK composites for bone implant applications. *Acta Biomaterialia*, **3**, 209-220.

30. Voyiadjis, G. Z., Samadi-Dooki, A., Malekmotiei, L. (2017) Nanoindentation of high performance semicrystalline polymers: A case study on PEEK. *Polymer Testing*, **61**, 57-64.

31. *Functional Polymer Blends*, Mittal, V. (ed.), CRC Press, USA (2012).

32. *Advances in Polyolefins*, Seymour, R. B., and Cheng, T. C. (eds.), Springer, USA (1987).

33. Schnell, H. (1956) Polycarbonate, eine gruppe neuartiger thermo-plastischer kunststoffe. herstellung und eigenschaften aroma-tischer polyester der kohlensäure. *Angewandte Chemie*, **68**(2), 633-640.
34. *Applied Polymer Science*, Craver, J. K., and Tess, R. W. (eds.), American Chemical Society, USA (1975).
35. De Leon, A. C., Chen, Q., Palaganas, N. B., Palaganas, J. O., Manapat, J., and Advincula, R. C. (2016) High performance polymer nanocomposites for additive manufacturing applications. *Reactive and Functional Polymers*, **103**, 141-155.
36. Carothers W. H. (1938) Diamine-dibasic Acid Salts, patent US2130947.
37. *Polymer Brushes*, Mittal, V. (ed.), CRC Press, USA (2012).
38. *Miniemulsion Polymerization Technology*, Mittal, V. (ed.), John Wiley, USA (2010).
39. Ligon, S. C., Liski, R., Stampfl, J., Gurr, M., and Mülhaupt, R. (2017) Polymers for 3D printing and customized additive manufacturing. *Chemical Reviews*, **117**, 10212-10290.
40. Lee, K. S., Kim, R. H., Yang, D. Y., and Park, S. H. (2008) Advances in 3D nano/microfabrication using two-photon initiated polymerization. *Progress in Polymer Science*, **33**, 631-681.
41. *Old World Labs*. Online: https://www.oldworldlabs.com [accessed 12th June 2019].
42. Mudge, R. P., and Wald, N.R. (2007) Laser engineered net shaping advances additive manufacturing and repair. *Welding Journal*, **86**, 44-48.
43. Singh, R. (2011) Process capability study of polyjet printing for plastic components. *Journal of Mechanical Science and Technology*, **25**, 1011-1015.
44. *Fast, Precise, Safe Prototype with FDM*. Online: http://sffsymposium.engr.utexas.edu/Manuscripts/1991/1991-15-Wales.pdf [accessed 19th June 2019].
45. Noorani, R. (2006) *Rapid Prototyping: Principles and Applications*, John Wiley & Sons, USA.
46. Gibson, I., Rosen, D., and Stucker, B. (2015) *Additive Manufacturing Technologies*, 2nd edition, Springer, USA.
47. Sciancaleporea, C., Moroni, F., Messoria, M., and Bondioli, F. (2017) Acrylate-based silver nanocomposite by simultaneous polymerization–reduction approach via 3D stereolithography. *Composites Communications*, **6**, 11-16.
48. Mota, C., Puppi, D., Dinucci, D., Gazzarri, M., and Chiellini, F. (2013) Additive manufacturing of star poly(ε-caprolactone) wet-spun scaffolds for bone tissue engineering applications. *Journal of Bioactive and Compatible Polymers*, **28**, 320-340.
49. Muth, J. T., Vogt, D. M., Truby, R. L., Mengüç, Y., Kolesky, D. B., Wood,

R. J., and Lewis, J. A. (2014) Embedded 3D printing of strain sensors within highly stretchable elastomers. *Advanced Materials*, **26**, 6307-6312.

50. Le, H. P. (1998) Progress and trends in ink-jet print technology. *Journal of Imaging Science and Technology*, **42**, 49-62.

51. Feygin, M., and Hsieh, B. (1991) Laminated Object Manufacturing (LOM): A Simpler Process. *1991 International Solid Freeform Fabrication Symposium.*

52. Kamrani, A. K., and Nasr, E. A. (2010) *Engineering Design and Rapid Prototyping*, Springer, USA.

53. Sachs, E., Cima, M., and Cornie, J. (1990) Three dimensional printing: rapid tooling and prototypes directly from a CAD model. *CIRP Annals*, **39**, 201-204.

54. Huang, S. H., Liu, P., Mokasdar, A., and Hou, L. (2013) Additive manufacturing and its societal impact: A literature review. *The International Journal of Advanced Manufacturing Technology*, **67**, 1191-1203.

55. Kim, G. D., and Oh, Y. T. (2008) A benchmark study on rapid prototyping processes and machines: quantitative comparisons of mechanical properties, accuracy, roughness, speed, and material cost. *Proceedings of the Institution of Mechanical Engineers, Part B*, **22**, 201-215.

56. Deckard, C., and Beaman, J. J. (1988) Process and control issues in selective laser sintering. *Sensors and Controls for Manufacturing*, USA, pp. 191-197.

57. Beaman, J. J., Barlow, J. W., Bourell, D. L., Crawford, R. H., Marcus, H. L., and McAlea, K. P. (1996) *Solid Freeform Fabrication: A New Direction in Manufacturing*, Springer, USA.

58. Murr, L. E., Gaytan, S. M., Ramirez, D. A., Martinez, E., Hernandez, J., Amato, K. N., Shindo, P. W., Medina, F. R., Wicker, R. B. (2012) Metal fabrication by additive manufacturing using laser and electron beam melting technologies. *Journal of Material Science and Technology*, **28**, 1-14.

59. Tang, H. H., Chiu, M. L., and Yen, H. C. (2011) Slurry based selective laser sintering of polymer-coated ceramic powders to fabricate high strength alumina parts. *Journal of European Ceramic Society*, **31**, 1383-1388.

60. Salentijn, G. I. J., Oome, P. E., Grajewski, M., and Verpoorte, E. (2017) Fused deposition modeling 3d printing for (bio)analytical device fabrication: procedures, materials, and applications. *Analytical Chemistry*, **89**, 7053-7061.

61. Pham, D. T., Ji, C. (2000) Design for stereolithography. *Proceedings of the Institution of Mechanical Engineers*, **214**, 635-640.

62. Melchels, F. P. W., Feijen, J., Grijpma, D. W. (2010) A review on stereolithography and its applications in biomedical engineering. *Bio-*

materials, **31**, 6121-6130.

63. Therriault, D., Shepherd, R. F., White, S. R., and Lewis, J. A. (2005) Fugitive inks for direct-write assembly of three-dimensional micro-vascular networks. *Advanced Materials*, **17**, 395-399.

64. Duoss, E. B., Weisgraber, T. H., Hearon, K., Zhu, C., Small, W., Metz, T. R., and Spadaccini, C. M. (2014) Three-dimensional printing of elastomeric, cellular architectures with negative stiffness. *Advanced Functional Materials*, **24**, 4905-4913.

65. Nelson, A., and Cosgrove, T. (2004) Dynamic light scattering studies of poly(ethylene oxide) adsorbed on laponite: Layer conformation and its effect on particle stability. *Langmuir*, **20**, 10382-10388.

66. Schmidt, G., Nakatani, A. I., and Han, C. C. (2002) Rheology and flow-birefringence from viscoelastic polymer-clay solutions. *Rheologica Acta*, **41**, 45-54.

67. Nelson, A., and Cosgrove, T. (2004) A small-angle neutron scattering study of adsorbed poly(ethylene oxide) on laponite. *Langmuir*, **20**, 2298-2304.

68. Mohanty, R. P., Suman, K., and Joshi, Y. M. (2017) In situ ion induced gelation of colloidal dispersion of Laponite: Relating microscopic interactions to macroscopic behavior. *Applied Clay Science*, **138**, 17-24.

69. Zulian, L., Ruzicka, B., and Ruocco, G. (2008) Influence of an adsorbing polymer on the aging dynamics of Laponite clay suspensions. *Philosophical Magazine*, **88**, 4213-4221.

70. Zulian, L., Marques, F. A. D., Emilitri, E., Ruocco, G., and Ruzicka, B. (2014) Dual aging behavior in a clay-polymer dispersion. *Soft Matter*, **10**, 4513-4521.

71. Peak, C. W., Stein, J., Gold, K. A., and Gaharwar, A. K. (2017) Nanoengineered colloidal inks for 3D bioprinting. *Langmuir*, **34**(3), 917-925.

72. Loffredo, F., Grimaldi, I. A., Mauro, A. D. G. D., Villani, F., Bizzarro, V., Nenna, G., D'Amato, R., and Minarini, C. (2011) Polyethylenimine/N-Doped titanium dioxide nanoparticlebased inks for ink-jet printing applications. *Journal of Applied Polymer Science*, **122**, 3630-3636.

73. Giardi, R., Porro, S., Chiolerio, A., Celasco, E., and Sangermano, M. (2013) Inkjet printed acrylic formulations based on UV-reduced graphene oxide nanocomposites. *Journal of Materials Science*, **48**, 1249-1255.

74. Arango, M. A. T., Andrade, A. S. V., Cipollone, D. T., Grant, L. O., Korakakis, D., and Sierros, K. A. (2016) Robotic deposition of TiO_2 films on flexible substrates from hybrid inks: Investigation of synthesis-processing-microstructure-photocatalytic relationships. *ACS Applied Materials & Interfaces*, **8**, 24659-24670.

75. Hayashi, K., Morii, H., Iwasaki, K., Horie, S., Horiishi, N., and Ichimura, K. (2007) Uniformed nano-downsizing of organic pigme-

nts through core-shell structuring. *Journal of Materials Chemistry*, **17**, 527-530.

76. Fu, S., Ding, L., Xu, C., and Wang, C. (2010) Properties of copper phthalocyanine blue encapsulated with a copolymer of styrene and maleic acid. *Journal of Applied Polymer Science*, **117**, 211-215.

77. Fu, S., Xu, C., Du, C., Tian, A., and Zhang, M. (2011) Encapsulation of C.I. Pigment blue 15:3 using a polymerizable dispersant via emulsion polymerization. *Colloids and Surfaces A*, **384**, 68-74.

78. Nguyen, D., Zondanos, H. S., Farrugia, J. M., Serelis, A. K., Such, C. H., and Hawkett, B. S. (2008) Pigment encapsulation by emulsion polymerization using macro-RAFT copolymers. *Langmuir*, **24**, 2140-2150.

79. Lelu, S., Novat, C., Graillat, C., Guyot, A., and Bourgeat-Lami, E. (2003) Encapsulation of an organic phthalocyanine blue pigment into polystyrene latex particles using a miniemulsion polymerization process. *Polymer International*, **52**, 542-547.

80. Steiert, N., and Landfester, K. (2007) Encapsulation of organic pigment particles via miniemulsion polymerization. *Macromolecular Materials and Engineering*, **292**, 1111-1125.

81. Fu, S. H., and Fang, K. J. (2007) Preparation of styrene-maleic acid copolymers and its application in encapsulated pigment red 122 dispersion. *Journal of Applied Polymer Science*, **105**, 317-321.

82. Yang, J., Wang, T. J., He, H., Wei, F., and Jin, Y. (2003) Particle size distribution and morphology of in situ suspension polymerized toner. *Industrial and Engineering Chemistry Research*, **42**, 5568-5575.

83. Kiatkamjornwong, S., and Pomsanam, P. (2003) Synthesis and characterization of styrenic-based polymerized toner and its composite for electrophotographic printing. *Journal of Applied Polymer Science*, **89**, 238-248.

84. Widiyandari, H., Iskandar, F., Hagura, N., and Okuyama, K. (2008) Preparation and characterization of nanopigment-poly(styrene-co-w-butyl acrylate-co-methacrylic acid) composite particles by high speed homogenization-assisted suspension polymerization. *Journal of Applied Polymer Science*, **108**, 1288-1297.

85. Qi, D., Zhang, R., Xu, L., Yuan, Y., and Lei, L. (2011) Preparation and characterization of organic pigment phthalocyanine blue microcapsules by in-situ micro-suspension polymerization. *Acta Polymerica Sinica*, **11**, 145-150.

86. Qi, D., Chen, Z., Yang, L., Cao, Z., and Wu, M. (2013) Improvement of dispersion and color effect of organic pigments in polymeric films via microencapsulation by the miniemulsion technique. *Advances in Materials Science and Engineering*, **2013**, doi: 10.1155/2013/790321.

87. Zeng, Q. H., Yu, A. B., Lu, G. Q., and Paul, D. R. (2005) Clay-based pol-

ymer nanocomposites: research and commercial development. *Journal of Nanoscience and Nanotechnology*, **5**, 1574-1592.

88. Abdelrazek, E. M., Elashmawi, I. S., Hezma, A. M., Rajeh, A., and Kamal, M. (2016) Effect of an encapsulate carbon nanotubes (CNTs) on structural and electrical properties of PU/PVC nanocomposites. *Physica B*, **502**, 48-55.

89. Elashmawi, I. S., Alatawi, N. S., and Elsayed, N. H. (2017) Preparation and characterization of polymer nanocomposites based on PVDF/PVC doped with graphene nanoparticles. *Results in Physics*, **7**, 636-640.

90. Jadhav, N., and Gelling, V. (2014) Titanium dioxide/conducting polymers composite pigments for corrosion protection of cold rolled steel. *Journal of Coatings Technology and Research*, **12**, 137-152.

91. Ghannam, L., Garay, H., Shanahan, M. E. R., Francüois, J., and Billon, L. (2005) A new pigment type: Colored diblock copolymer-mica composites. *Chemistry of Materials*, **17**, 3837-3843.

10

Biopolymer Coatings

10.1 Introduction

Metallic materials and structures exposed to aggressive environments during application undergo deterioration and premature failure because of the processes of corrosion. Application of polymeric coatings is the standard procedure widely employed to control the corrosive destruction of the metal structures [1-3]. However, many of the conventional coating systems lead to several environment and health related issues such as emission of volatile organic solvents and difficulty in recycling or waste disposal of resins at the end of their usage. Thus, to overcome these issues, there is an increasing demand for new "green or bio" coating technologies based on waterborne or high solid content coating formulations for the corrosion protection of metallic structures [4-6]. In addition to overcoming the environmental issues, the trend for biopolymer coatings has also evolved from the realization that the fossil resources used for the conventional organic coatings are inherently finite. Thus, it is vital to develop effective biopolymer coating systems by using film forming polymers originated from bio-sources. This approach will not only cut down the growing prices of raw materials, but would also generate environmentally friendly coatings, and solve the waste disposal problems of resin materials. Cellulose, chitosan, starch, polylactide, vegetable oils, etc., are some of the bio-polymers extensively developed for the coating applications due to their inherent properties of biodegradability and low toxicity [7,8]. In addition to the corrosion resistant coatings and paints, biopolymers can also be formulated as corrosion resistant pigments, alloys and composites. These macromolecules are rich in functional groups which provide multiple adsorption sites suitable to form strong complexes with metal ions and, thus, protect them from corrosion [9-11]. Their performance can be further improved by suitable chemical modifications or by the use of reinforcing materials like nanoparticles. This chapter presents an overview on the development of various biopolymer coatings and their ability in

Gisha E. Luckachan and Vikas Mittal, Khalifa University of Science and Technology, Abu Dhabi, UAE

protecting metal substrates in different corrosive environments in different process industries.

10.2 Cellulose

Cellulose, one of the most abundant biopolymers functionalized naturally with hydroxyl and amino groups, is commonly utilized to inhibit metal corrosion in acid media. Derivatives of cellulose such as carboxymethyl cellulose [12,13], hydroxypropyl cellulose [14], ethyl hydroxyethyl cellulose [15], hydroxypropyl methyl cellulose (HPMC) [16] and aminated hydroxyethyl cellulose [17] have been widely used as corrosion inhibitors. A mixed type of inhibition (i.e. inhibiting both anodic and cathodic processes of corrosion) was also reported for cellulose derivatives, especially for hydroxyethyl cellulose (HEC) in acid solutions [17]. Inhibition occurred by the coordinate bonding of the polar functional groups of the HEC polymer with the metal surface and the surface coverage by the bulky cyclic ring structures of cellulose adsorbed on it. Thus, HEC adsorption on the metal surface occurs through multiple OH functional groups and the aromatic ring structures. An enhanced inhibitive effect of HEC was reported by Arukulam *et al.* [17] by the incorporation of an inhibitor containing halide additive (KI). The iodide ions stabilized the cellulose adsorption on the metal surface, which occurred by the coulombic interaction between HEC$^+$ and I$^-$ ions. Firstly, the iodide ions chemisorbed strongly on the metal surface, and HEC$^+$ adsorbed subsequently, which led to higher extent of surface coverage and, hence, greater corrosion inhibition. Potentiodynamic polarization measurements exhibited a mixed type corrosion inhibition using HEC+KI combination. The anodic and cathodic branches of Tafel plots were observed to shift towards lower values of corrosion current density, as compared to the plots corresponding to HEC alone, thus, confirming superior performance.

In another study, Mobin *et al.* [18] employed the mixing of a minimal concentration of surfactants, triton X 100 (TX), cetyl pyridinium chloride (CPC) and sodium dodecyl sulfate (SDS) with hydroxyethyl cellulose to enhance the inhibitive effect on A1020 carbon steel corrosion in acid solutions. The concentration of surfactants was kept very low to maintain the eco-friendly nature of HEC. The enhanced inhibition of HEC in the presence of the surfactant occurred by a protective film formation on the metal surface by the interaction between the HEC and the surfactant. In acid solution, Cl$^-$ ions present in the electrolyte solution adsorbed on the metal surface to which HEC molecules, protonated in acid solution, attached by electrostatic interaction. Surfactant was then attached to HEC which led to a large surface coverage, thus, providing a better corrosion protection than pure HEC molecules. Each surfactant interacted differently with HEC in acid solution. In 1M HCl, SDS and TX were suggested to interact electrostatically with HEC, whereas CPC binding to the HEC in 1M HCl was opined occur through hydrophobic group i.e. an attractive lipophilic interaction between the surfactant tail and polymer chain. Moreover, the polymer backbone exhibited an expansion by the repletion of charged groups on ionic surfactant bound to HEC. This expansion helped the polymer molecules to occupy more surface area on carbon steel surface and, thus, protect it from aggressive medium.

HPMC has also been reported to exhibit a mixed-type corrosion inhibition in 0.5M H_2SO_4 and 1M HCl solutions and the extent of inhibition was influenced by the inhibitor concentration, temperature of the media and time of exposure [19,20]. The inhibitive effect of HPMC was by the formation of protective layer on the metal surface through the hydroxypropyl end and the bulky aromatic ring of the molecule. It was reported that in the absence of corrosion, a physical blocking of active sites and in the presence of corrosion, a geometric blocking of active sites on the mild steel occurred due to HPMC molecules. The compound carboxy methyl cellulose was also reported as good inhibitor for mild steel corrosion in aqueous environments [13,21,22]. Another derivative of cellulose, carboxymethyl cellulose-1-hydroxyethane-1,1- diphosphonic acid, was investigated by Rajendran *et al.* [23] for the inhibitive behavior of carbon steel corrosion. An inhibition efficiency of 83.34% was observed for sodium carboxymethyl cellulose (Na-CMC) at a concentration of 5 mg/L at 20 °C [24]. Similar to other cellulose derivatives, Na-CMC also provided protection by the adsorption of inhibitor molecules at metal/solution interface [25]. The calculations of thermodynamic parameters indicated that a chemical adsorption of carboxyl group of Na-CMC occurred on the metal surface following a Langmuir's adsorption isotherm. Thus, a large number of cellulose derivatives have been confirmed as effective corrosion inhibitors, which underlines the usefulness of bio-based polymers for advanced applications.

Yang *et al.* [26] reported transparent cellulose films prepared from aqueous alkali (NaOH or LiOH)/urea (AU) solutions. The films exhibited high oxygen barrier properties in comparison with conventional cellophane films. As shown in Figure 10.1, the cellulose films also exhibited significantly lower water vapor permeability as compared to cellophane over the entire range of relative humidity, thus, further confirming the potential of these cellulose coatings to be useful alternatives to the conventional coating materials. In another study, Qi *et al.* [27] fabricated composite coatings through layer-by-layer assembly of cellulose and chitin nanofibrils. The layer by layer coatings on PET films exhibited high transparency, surface hydrophilicity, flexibility and nanoporous structures. With further methodological developments, the authors suggested the potential use of developed coatings systems in a wide variety of applications. Figure 10.2 shows the AFM height images of the surfaces of (a) 20- and (b) 21-layered films. Cellulose acetate (CA) films combined with nanoparticles have also been reported as an effective treatment for corrosion protection of metals [28-30]. Tamborim *et al.* [31] studied the corrosion inhibitive effect of cellulose acetate coating doped with amoxicillin on AA2024-T3 aluminum alloy in 0.05 M NaCl solution. An effective protection was reported for CA films doped with 2000 ppm of amoxicillin, which occurred due to its ability to form different complexes with aluminium ions [32].

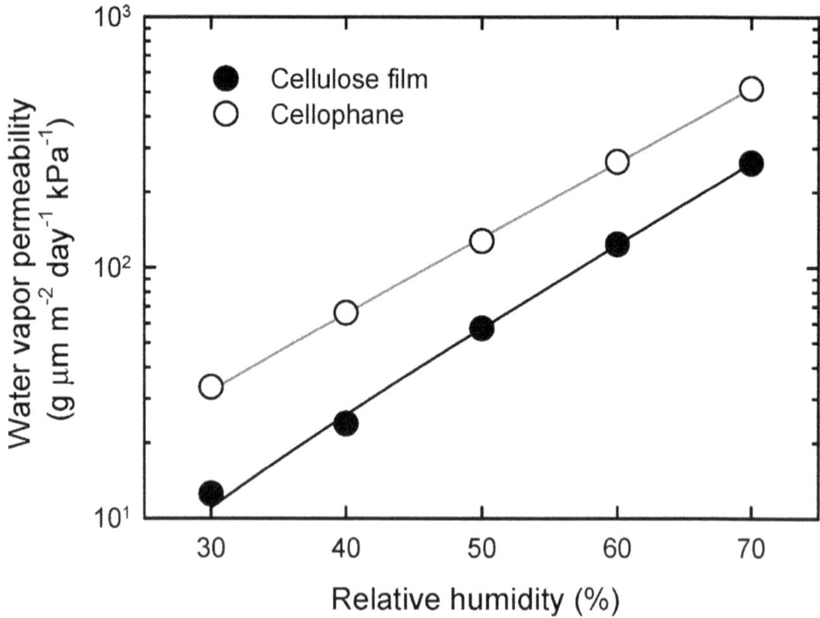

Figure 10.1 Effect of relative humidity on water vapor permeability of AU cellulose and cellophane. Reproduced from Reference 26 with permission from American Chemical Society.

Yabuki *et al.* [33,34] used cellulose nanofiber incorporated with corrosion inhibitor (calcium nitrite) to prepare self-healing polymer coatings. The coatings were scratched for polarization measurements and a higher resistance was observed for nanofiber/corrosion inhibitor coating as compared to the coating containing only corrosion inhibitor. The method represented an ideal methodology for the fabrication of biopolymer based self-healing coatings because of the network structure of cellulose nanofiber and distribution of the healing agent uniformly over the coating. Nanofibers provided a pathway for the release of healing agent through the coating. It was confirmed in the cross-section of the polymer scratch where empty holes of sizes equivalent to the diameter of cellulose nanofibers were observed after the corrosion test. Since cellulose nanofibers have several OH functional groups on their backbone chain, a pH controlled

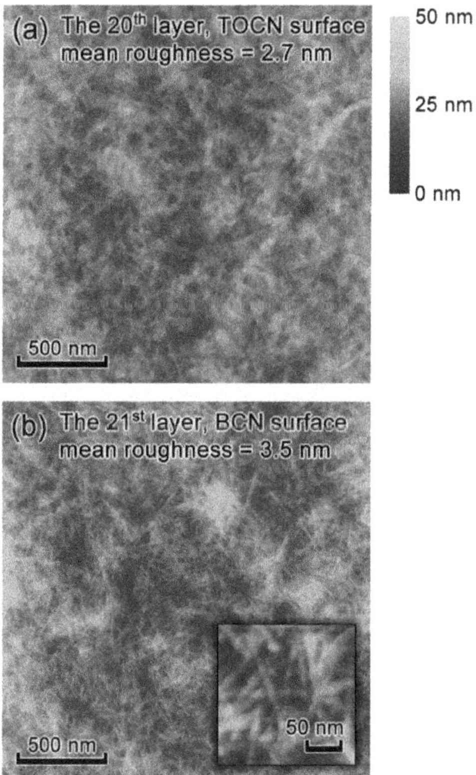

Figure 10.2 AFM height images of the surface of (a) 20- and (b) 21-layered films; TOCN: cellulose nanofibrils and BCN: chitin nanofibrils. Reproduced from Reference 27 with permission from American Chemical Society.

adsorption and desorption from the nanofiber chain is possible in the suitable environments. The authors also studied the behavior of nanofibers using oleic acid (OA) as corrosion inhibitor and sodium hydroxide as pH adjuster [35]. The polarization resistance of the scratched surfaces of nanofiber-oleic acid coatings were two orders of magnitude higher than the plain coating in a pH 11.4 solution. The protection occurred by the formation of a stable film on the metal surface by the interaction of oleic acid released through the nanofiber pathways with metal ions. The scratch exposed the metal surface towards the electrolyte that activated the anodic processes forming Fe^{2+} ions. The cathodic process occurring near the contact between the coating and the substrate formed OH^- ions which diffused through the coating, and resulted in an increase of local pH and promoted the release of oleic acid from the surface of nanofibers. The triggering action for such a self-healing process was the cathodic reaction on the metal surface.

10.3 Chitosan

Chitosan (CS), a derivative of polysaccharide chitin, is second to cellulose in natural abundance [36,37]. Its production is estimated around one billion tons per year. CS is a linear biopolymer of poly(d-glucosamine) connected to poly(*N*-acetyl-d-glucosamine) through a beta 1→4 linkage. The amine functional groups create positive charge density on the chitosan chain in slightly acidic medium [38]. Such a cationic property together with its biocompatibility make CS widely useful in biomedical applications. Its film forming and gelation properties enable it to be used as a barrier coating for the corrosion protection of metals as well. In addition, lone pairs of electrons on the O and N atoms of hydroxyl and amine functional groups of chitosan can interact ionically with the metal surface, so it is considered as a natural corrosion inhibitor [39,40,]. El-Haddad [41] reported 95% efficiency in copper corrosion inhibition by chitosan in acid medium (0.5 M HCl). Similar to cellulose, chitosan also acted as a mixed-type inhibitor which reduced both anodic and cathodic processes of copper corrosion. It was observed from the potentiodynamic measurements that E_{corr} shifted to cathodic side and the corrosion current density decreased with increase in chitosan concentration. The inhibition mechanism was driven by the electrostatic interaction of metal ions with Cl^- and positively charged chitosan in the acid medium. Copper surface, positively charged in acid solution, got adsorbed by Cl^-

ions from the electrolyte, thus, keeping the metal surface more negatively charged in the solution. The positively charged chitosan molecules subsequently attach to the Cl⁻ ions on the metal forming a thin barrier layer [42]. In addition to the ionic adsorption, the distribution of electrons between the free heteroatoms of chitosan chain with vacant *d* orbitals of iron results in a chemical adsorption of cationic and neutral molecules on the metal surface. In addition to the pure chitosan, coatings of chitosan derivatives like acetyl thiourea chitosan and naturally available carboxymethyl chitosan also exhibited an efficient mild steel corrosion inhibition in acid solution [43,44]. Mild steel corrosion inhibition in sodium chloride solution was reported by Mohamed and Fekry using chitosan-crotonaldehyde Schiff's base [45]. The barrier property of chitosan coatings along with physical and mechanical properties could be enhanced by crosslinking reactions. Glutaraldehyde [46], sulfuric acid [47], epoxy [48] and dialdehyde starch [49] are the commonly used crosslinking agents for chitosan. Luckachan and Mittal [50] reported anti-corrosive chitosan coatings for mild steel protection using gluatraldehyde as crosslinking agent along with a hydrophobic poly(vinylbutyral) (PVB) overcoating. Figure 10.3 shows the optical photographs of steel surfaces coated with PVB_Ch/x%Glu_PVB layers containing different amount of glutaraldehyde after 24 h EIS measurent and film removal. Metal surface beneath the coating was observed to be uniformly covered by a thin film, which appeared in the SEM image as uniform less-porous thick precipitate. Raman spectra of the area exhibited characteristic bands of Fe_3O_4 and γ-Fe_2O_3 oxides. Hoeevr, the SEM image of the bare steel after corrosion analysis exhibited that corroison products were agglomerated. It indicated that chitosan coating prevented the agglomeration of passive iron oxide layers on the metal surface and, thus, helped to cover the surface uniformly. The chitosan stabilized iron oxide layer separated the corosive agents from reaching the metal surface and, thus, prevented further corrosion of the metal beneath the oxide layer. Such a protection is obtained by the chelating effect of chitosan with iron ions. These findings indicated that if the structure of chitosan was suitably modified, the obtained coating formulations represent high potential of large scale use in process industries.

Another approach to enhance the properties of chitosan is the addition of inorganic particles, which significantly enhances the oxygen barrier and adhesion strength. In a recent study, Luckachan and Mittal [51] (Figure 10.4) generated stable chitosan-silica based coatings

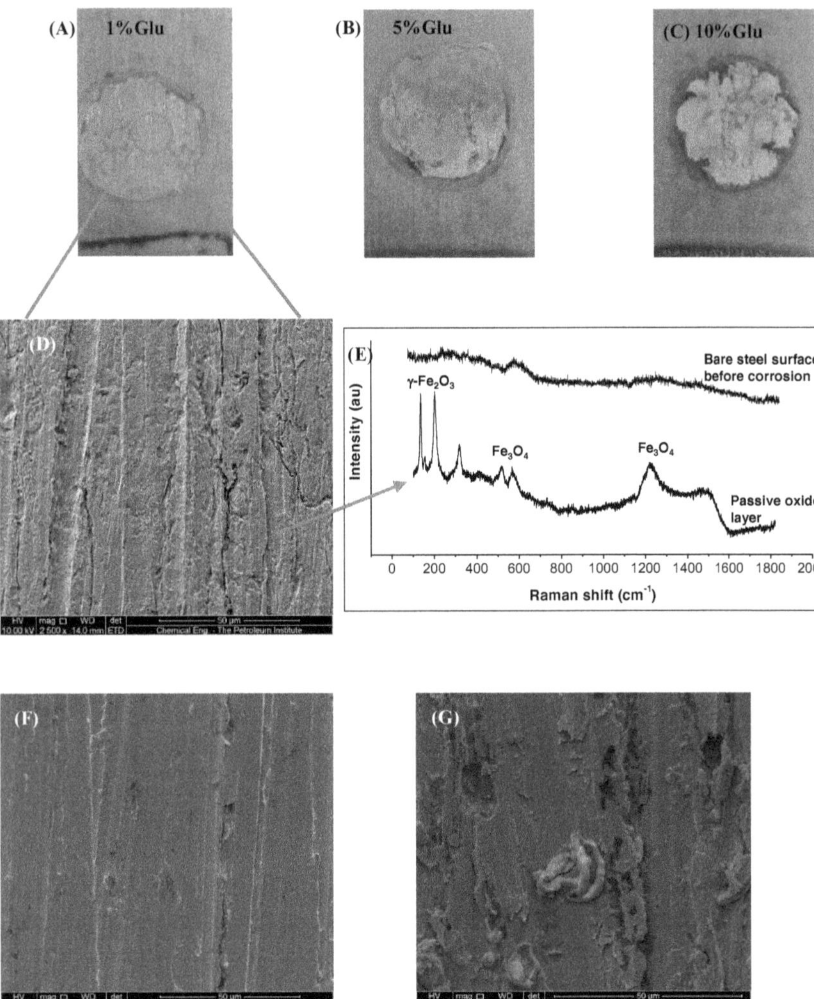

Figure 10.3 Optical photographs of PVB_Ch/1%Glu_PVB coated steel surface containing (A) 1%Glu, (B) 5%Glu and (C) 10%Glu after 24 h EIS in 0.3 M salt solution and film removal. Diameter of the circle was 1 cm². (D) SEM image and (E) Raman spectra of the area marked with red circle on optical photograph of steel surface taken after PVB_Ch/1%Glu_PVB coating removal. SEM images of bare steel (F) before and (G) after 24 h EIS measurement in 0.3 M salt solution for comparison of the corrosion performance with the composite coating. Reproduced from Reference 50 with permission from Springer.

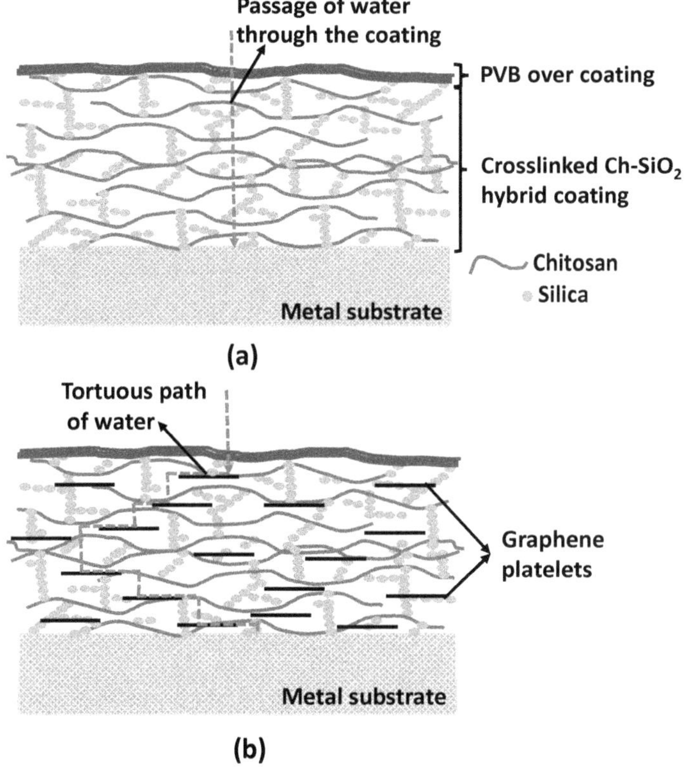

Figure 10.4 Schematic representation of corrosion protection provided by Ch-SiO$_2$ hybrid coatings in the absence (a) and presence (b) of graphene platelets. Reproduced from Reference 51.

on carbon steel substrates [51]. Effect of chitosan content, cross-linking with silica, poly(vinyl butyral) (PVB) over-coat as well as graphene reinforcement on the anti-corrosion performance of the coatings in an aggressive environment was studied. An excellent corrosion resistance was obtained by the application of a PVB over-coating on to Ch-SiO$_2$ hybrid coating, even though the overall coating thickness was still very low (5-6 μm). Uniformly distributed graphene platelets contributed to the significantly enhanced corrosion resistance by acting as water barrier, which was reflected in the lower coating capacitance and I_{corr}. The schematic representation of the corrosion protection in the absence and presence of graphene platelets is presented in Figure 10.4.

Chitosan modified with β-cyclodextrin was also reported by Liu *et al.* [52] as a green inhibitor to inhibit the carbon steel corrosion in 0.5

M hydrochloric acid solutions. The inhibition efficiency was observed to increase with inhibitor concentration (Figure 10.5), with the maximum inhibition efficiency of 96.02%. Inhibition process of β-cyclodextrin modified natural chitosan followed the Langmuir adsorption isotherm, where metal surface was covered and protected by the physical and chemical adsorption of chitosan.

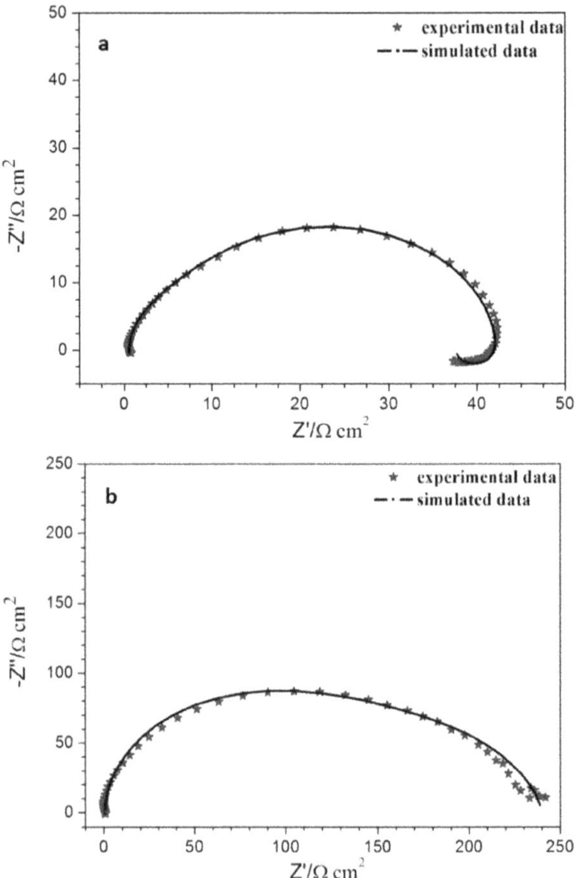

Figure 10.5 Simulated and experimental impedance spectra in 0.5 M HCl solution: (a) blank solution; (b) in the presence of 80 ppm inhibitor. Reproduced from Reference 52 with permission of American Chemical Society.

In another study, Hoagland and Parris [53] casted chitosan/lactic acid films on pectin films using either glycerol or lactic acid as plasticizer. The storage modulus of the chitosan/pectin laminated films was observed to be significantly higher than the pure chitosan films (Figure 10.6).

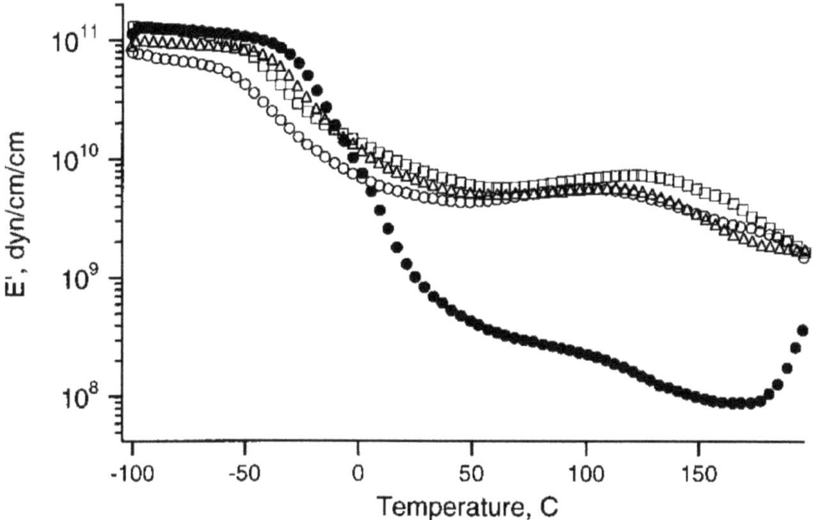

Figure 10.6 Storage modulus for pectin (P)/lactic acid (LA) film (P:LA 2:1, open circles), for chitosan (C)/lactic acid film (C:LA 2:2, solid circles), and for laminates (P:LA:C:LA 2:1:1:1, squares, and P:LA:C:LA 1:0.5:1:1, triangles). Reproduced from Reference 53 with permission from American Chemical Society.

The affinity of chitosan towards water is an undesirable property for its application as the corrosion mitigating coating films. Chitosan changes to a hydrogel by absorbing moisture from the environment [54,55]. It causes the easy degradation of the film and, thus, leads to the complete failure of chitosan as a protective barrier coating. Several works have reported the modification of chitosan structure using organic or bio-organic materials to reduce its water affinity. Sugama and Milian-Jimenez [56] reported the corn-starch derived dextrin modified chitosan for the protection of aluminium. Coating formulations were prepared by mixing chitosan dissolved in HCl with corn starch-derived dextrin (DEX) containing Ce nitrate as oxidizing agent. The mixture was dip-coated on aluminium surface, followed by heating to obtain a solid film. Under this condition, DEX oxidized to two fragmental products of 3,4-dihydroxybutanoic acid and glycolic acid. These fragments formed Ce-complexed carboxylate by reacting with Ce in the Ce nitrate. Ce-bridged carboxylate fragments had a strong affinity with the primary amine in chitosan chain and created a secondary amide linkage that kept the DEX grafted onto the chitosan chain. DEX grafting on CS resulted in the formation of uniform defect-free coatings with less susceptibility to moisture, along with reduced permeability to corrosive species through the coating. CS/DEX coating obtained from a formulation ratio of 70/30 withstood corrosion in salt-spray analysis for over 720 h. In addition, 70/30CS/DEX coating exhibited pore resistance an order of magnitude higher than that of single CS and DEX coatings. These results indicated lower permeability and higher barrier protection of DEX modified chitosan coatings on aluminium surface. Kumar and Buchheit [57] modified chitosan structure using epoxy functional silane and vanadate for the corrosion resistant coating for Al 2024 T3 alloy. Adhesive strength of chitosan coatings increased significantly by the addition of 1% GPTS ((3-glycidoxypropyl)-trimethoxysilane)-vanadate in the coating formulation to an extent which was comparable to the adhesive strength of pure epoxy coatings. A controlled release of vanadate from the chitosan coating was obtained by proper pH adjustment that provided a self-healing nature to these coatings for corrosion protection. The highest release of vanadate occurred in pH 10 solution and a negligible release was observed in pH 3 solution. The maximum adsorption of vanadate was in the pH 3-5 range because the amine functional groups of chitosan getting protonated in the acidic range. Hence, the adsorption and desorption of vanadate onto chitosan was observed to be reversible. CS-1%GPTS-vanadate coatings remained for 450 h in the salt spray chamber without any indication of corrosion initiation.

Another approach to solve the water affinity issue of chitosan is the modification with negatively charged polyacid electrolytes, since chitosan exists as a cationic polybase in acidic media. Sugama and Cook [58] attempted this approach by molecular modification of chitosan with poly(itaconic acid) (PIA). PIA modified CS coatings were prepared by dip-coating CS-PIA solution in HCl on aluminum substrate followed by heating at 200 °C for over 120 min. Under such high temperature conditions, the primary amine groups in CS interacted strongly with the carboxylic acid groups of poly(itaconic acid) and formed a secondary amide linkage which grafted the PI polymer onto the chitosan backbone and crosslinked between the chitosan chains. Also, PIA exhibited a direct affinity with aluminium surface by forming a –COO-Al linkage by the reaction of the COOH groups in PIA with OH groups of aluminium surface. Such an interfacial interaction of PIA with metal surface along with the crosslinked structure of chitosan enhanced the barrier protection by reducing moisture sensitivity and infiltration of corrosive species through the coating. CS/PIA coating formulation of 80/20 ratio possessed the aforementioned properties required for a corrosion protecting barrier coating and provided protection to aluminium metal for over 694 h in salt spray chamber.

Crosslinking chitosan with metal ions is a different approach used to enhance the barrier properties of chitosan coatings. The lone pairs of electrons on amine and hydroxyl functional chitosan can involve in coordination with the transition metal ions. Such an interaction occurs at neutral pH, as at low pH the protonated amine decreases the affinity of chitosan to chelate with metal ions [59]. This coordination mechanism was explained by so-called bridge and pendant models. In the bridge model, the metal ion is bound to the nitrogen atoms of either same or different chitosan chains. In the pendant model, the metallic ions are bound with the amino groups as a pendant [60-62]. Copper incorporated chitosan films were reported by Lundvall *et al.* [63], for providing protection to AA-2024-T3 aluminium alloy in corrosive environment. Chitosan dissolved in acetic acid solution was coated first on anodized surface of aluminium. The sample were dipped in copper salt of acetate for 24 h to incorporate copper ions in between the chitosan chains. An enhanced protection of chitosan-Cu coatings over pure chitosan coating was observed in borax solution, which was confirmed by a constant high open circuit potential. Changes in the salt concentration didn't change the performance of the coatings, but pH of the copper salt solution affected significantly the copper incorporation into the chitosan structure. An optimum pH of 4 favored coordination of copper ions onto chitosan amine groups and, thus, crosslinked the chitosan chains. Such a crosslinked structure enhanced the barrier protection of chitosan coatings by decreasing its permeability towards the moisture and corrosive species.

Recently, several works have been reported on the use of chitosan as a reservoir of corrosion inhibitor for the controlled release at required areas of the coatings [64-66]. Carneiro *et al.* and Wang *et al.* [65,66] reported chitosan coatings loaded with organic corrosion inhibitor for the protection of aluminum alloys. Carneiro *et al.* [67] used chitosan solution loaded with 10 wt% of mercaptobenzotriazol (MBT) for the corrosion protection of AA2024 alloy. The main idea of testing CS-MBT was to use such thin biopolymer based coatings for a temporary corrosion protection and active protection of zones where barrier organic coatings cannot be applied. A localized nature of corrosion is usually observed on aluminum alloy 2024 in NaCl solutions within a neutral pH range. However, the local anodic and cathodic currents generated at the metal surface in the micrometer range can lead to considerable local pH changes, highlighting the importance of having a polymeric matrix responding to pH changes and an inhibitor which can suppress both anodic and cathodic corrosion processes. The release of MBT from CS is mainly based on the solubility of chitosan and MBT in different pH conditions. In acidic conditions, the solubility of chitosan allows partial dissolution of entrapped MBT, whereas in alkaline conditions, the solubility of MBT is the dominant parameter as chitosan is insoluble at this condition. In distilled water and NaCl solution, the range of pH is neutral and the combination of low solubility of CS and low solubility of MBT was accounted for a limited release of MBT. CS-MBT coated plates exhibited a clear and unchanged appearance after one week of immersion in 0.05 M sodium chloride solution, indicating the diminishing of corrosion processes by the inhibitor released from the chitosan matrix. Though incorporation of MBT into CS structure was beneficial in stopping the corrosion of aluminium alloy over a period longer than one week, it lacked the barrier nature because of its affinity towards moisture. Therefore, its use as a potential alternative to full coating systems is still not well established. The authors also attempted to solve this issue by functionalizing chitosan with hydrophobic entities like poly(ethylene-*alt*-maleic anhydride) (PEMA) and poly(maleic anhydride-*alt*-1-octadecene) (POMA) [65]. The aliphatic chains were grafted onto chitosan structure by reacting amine groups of CS with maleic anhydride functionality present in the aliphatic polymer chains. Hydrophobic nature of the coating was further enhanced by crosslinking with glutaraldehyde. It was observed by the water contact angle measurements that contact angle of chitosan increased from 69° ± 5° to 78° ± 6° by the loading of 10 wt% MBT. It was further increased to 137° ± 3° and 141° ± 6, by the grafting of PEMA and POMA respectively on to chitosan chains. Carneiro *et al.* [68] also incorporated Ce^{3+} ions into chitosan chains after chemical functionalization with 2,2,3,3-tetrafluoropropyl ether (GTFE). The coatings imparted an active corrosion protection by increasing the oxide layer resistance on the metal surface in presence of Ce^{3+} cations released from the chitosan matrix. However, the coatings didn't achieve the recommended barrier effect for high performance applications even after chitosan modification. Therefore, such coatings are anticipated to be a better alternative to synthetic polymer coatings for short term applications and temporary protection. In another study, Hassannejad and Nouri [69] studied the self-healing behavior of chitosan-cerium ion (CE) nanocomposite coatings on AA5083-H321 prepared by incorporating CeO_2 in chitosan chain. The authors reported extensively increased corrosion resistance with increase of cerium ions concentration and maximum efficiency was attained at 5 g.l^{-1} CeO_2 nanoparticles in the CS-CE formulation. The inhibitive efficiency of chitosan coating was attributed to the diffusion of cerium ions to the localized defects. As coating was placed in the corrosive solution, the cerium ion inhibited corrosion by creating thin protective layer of cerium oxide on the metal surface by changing the valency of Ce from +3 to +4. It indicated the self-healing nature of the coating, which was further confirmed from the high impedance values of CeO_2-CS-Ce (5 mM) coatings after a 120 h of exposure to the aggressive solution.

Chitosan is a suitable material for bio-corrosion protection because of its anti-bacterial property [70]. Bao *et al.* [71] studied this by using chitosan coatings on copper substrate to prevent the microbiologically influenced corrosion (MIC). Here, chitosan played a dual function, one as a carrier of corrosion inhibitor and second as a bacteriostatic matrix. In order to fabricate the chitosan coatings on copper substrate, a monolayer of alkane thiols was prepared firstly by immersing the pretreated copper surfaces in MUA (11-mercaptoundecanoic acid) solution of ethanol (5 mM). Chitosan-corrosion inhibitor (mercaptobenzotriazol (MBT)) formulation was then dip-coated onto the MUA modified copper surface. Corrosion analysis of these coatings showed a mixed type corrosion inhibition mechanism which decreased both cathodic and anodic current densities along with a positive shift of E_{corr} compared with the MUA and CS/MUA coatings. The authors reported a stable water insoluble product [Cu(I)-MBT]$_{ads}$ on the copper surface which was formed by the interaction of released MBT from chitosan coating with copper ions. Addition of inhibitor molecules directly to the corrosive environment is the generally used method for copper protection. The incorporation of inhibitor in the chitosan chains would be a beneficial alternative, as it minimizes the amount of inhibitor required for copper protection. In addition, corrosion protection performance of MBT-CS/MUA coating indicated effective anti-bacterial features as well. Copper coupons and MBT-CS/MUA coatings were exposed to sulphate reducing bacteria (SRB) culture media for 6 h. Corrosion was clearly observed in the FESEM images of the copper coupons with a rough and non-uniform surfaces along with the appearance of SRB bacteria growth. Such a settlement of SRB bacteria was not observed on the MBT-CS/MUA coated copper substrate indicating the protection of copper by the combined effect of MBT and chitosan.

Several studies have reported the chitosan nanocomposite coatings where the hydrophobicity and corrosion resistance of chitosan coatings were enhanced by the incorporation of suitable nano-fillers. Silica, graphene and graphene derivatives are the commonly employed nanofillers. Fayyad *et al.* [72] reported an enhanced hydrophobicity of chitosan coatings by incorporating graphene oxide (GO). A contact angle of 103.2° was obtained for CS/GO coatings compared with 88.4° of the pure chitosan coating. An oleic acid treatment of CS/GO films increased the contact angle further to 148°, which occurred due to the coverage of surface with bulky chains of oleic acid molecules. An excellent anti-corrosive nature occurred in CS/GO-OA coatings by the combined effect of GO and oleic acid moieties. GO reduced the permeability of water and oxygen through the coating by increasing tortuous path of diffusion and oleic acid reduced the hydrophilicity of the coating. Thus, the nano-filler reinforcement would be an effective method in the future studies to obtain biopolymer coatings for wide applications in the process industry.

10.4 Starch

Starch is a biodegradable polysaccharide composed of amylose (linear) and amylopectin (branched) molecules [73]. It is available at low cost in large amounts from agricultural crops. Since it is biodegradable starch, it is widely used in pharmaceutical, paper making, food preparation and textile applications. It is also considered as a natural corrosion inhibitor, however, its use as corrosion inhibitor is not well reported in the literature. Corn starch, potato, tapioca and wheat are commonly used refined starches. Rosliza *et al.* [74] examined the role of tapioca starch in improving the corrosion resistance of AA6061 alloy in seawater. Similar to other polysaccharides, tapioca starch also exhibited a mixed type corrosion inhibition mechanism with a little better control on the anodic reactions. A positive shift of corrosion potential was observed in the polarization curves along with a downward shift of anodic and cathodic branches of the curve with the addition of tapioca starch. The protection occurred by the formation of a thin film on the metal surface and, therefore, it was considered as a filming corrosion inhibitor which diminished the corrosion by creating a barrier between metal and the corrosive environment [75,76]. Effect of starch in preventing pitting corrosion of steel was studied by Abd El Haleem *et al.* [77] in aqueous environment. Bello *et al.* [78,79] used modified starch to protect carbon steel in alkaline conditions. Both activated starch (AS) and carboxymethylated starch (CMS) were analyzed for corrosion inhibition and the performance was observed to depend on the type and amount of end functional groups on the main chain of the polymer. CMS contained both carboxylate (–COO–) and alkoxy (–CO–) groups and AS contained alkoxy (–CO–) groups. AS had a strong affinity for ferrous ions which were attributed to a better anti-corrosion performance of AS over CMS. A thick inhibitive layer formed by the adsorption of AS on the metal surface was observed in the AFM images after 24 h of immersion showing the role of starch in providing steel corrosion protection. Sugama and DuVall [73] prepared a modified potato starch coating to inhibit the corrosion of aluminum substrates. Potato starch (PS) was modified by the opening of glycosidic rings with polyorganosiloxane (POS) grafting and then dip-coated on aluminum surface. A curing condition of 200 °C was employed to improve the hydrophobic characteristics of PS coating. The POS-grafted PS coating films deposited on the Al surface remained without any visible corrosion for over 288 h in salt spray analysis, confirming the barrier of the coating to moisture and corrosive species. The coating impedance also improved by a magnitude of two orders compared with the bare substrate. Though the above-mentioned polysaccharides have the ability to inhibit metal corrosion, they are generally limited in such applications because of their biodegradability and moisture sensitivity, unless the structure is suitably modified. As these are ecofriendly with low cost, thus, the coatings from these polymers have the potential to develop as suitable alternatives to the petroleum based polymer coatings, especially for short term applications.

10.5 Polylactide

Poly(lactic acid) or polylactide (PLA) is a biodegradable linear aliphatic thermoplastic polyester. It is widely produced from renewable resources like potato, corn and sugar beet, etc. It is mainly used for biomedical applications owing to its biodegradability, biocompatibility, mechanical properties and thermoplastic processability [80-82]. It has excellent oil resistance and gas impermeability and can also be processed easily. Apart from medical applications, it is used widely in the areas of packaging, textile, pharmaceutical uses and coatings [83-85]. Recently, PLA has attracted significant attention in the area of protective coatings of exterior surfaces of building materials because of its hydrophobic nature. It is useful in protecting buildings, especially historic monuments exposed to vigorous outdoor conditions. Svagan *et al.* [86] reported transparent films based on PLA and montmorillonite, as shown schematically in Figure 10.7. To achieve this, layers of montmorillonite and chitosan were generated on PLA films by a layer by layer addition method. The hybrid films exhibited superior properties as oxygen permeation of PLA was reduced respectively by 99 and 96%, at 20 and 50% RH, when 70 bilayers were deposited. Such systems represent potentially useful alternatives to achieve high performance coatings for industrial applications. Zeng *et al.* [87] studied the characteristics of a microarc oxidation (MAO) and poly(L-lactic acid) (PLLA) composite coating, which was fabricated on Mg–1Li–1Ca alloy by dip-coating and freeze-drying (Figure 10.8). MAO/PLLA composite coatings increased the corrosion resistance significantly that was attributed to an increase of E_{corr} from −1.66 V to −1.44 V and a decrease of I_{corr} by approx. 2 orders of magnitude.

Ocak *et al.* [88] reported the hydrophobicity on marble surfaces by the application of PLA coatings. It also diminished the formation of gypsum on marble surfaces when exposed to polluted environment. The authors noticed an increase in hydrophobicity of PLA coatings by the addition of nanoclay in the coating formulation. It enhanced the life time of the coating and the resistance of marble surfaces towards water and atmospheric pollutants [89]. A hydrophobic fluoro-functionalized PLA polymer suitable for stone protection was also prepared [90-92]. The fluorine atoms present in the PLA coating improved hydrophobicity as well as the resistance of coating to chemical and physical degradation agents. Pedna *et al.* [93] prepared highly hydrophobic coatings for buildings using SiO_2-fluorinated PLA bionanocomposites. The silica particles created roughness and the incorporated fluorine lowered the surface energy of the coating. The combined action of these two effects increased the hydrophobicity of the coating, which was confirmed from the water contact angle of about 140° on PLA coated building marble. To the best of knowledge, only a limited work on the use of PLA as a corrosion resistant coating for protecting the metallic structures has been reported so far in the literature.

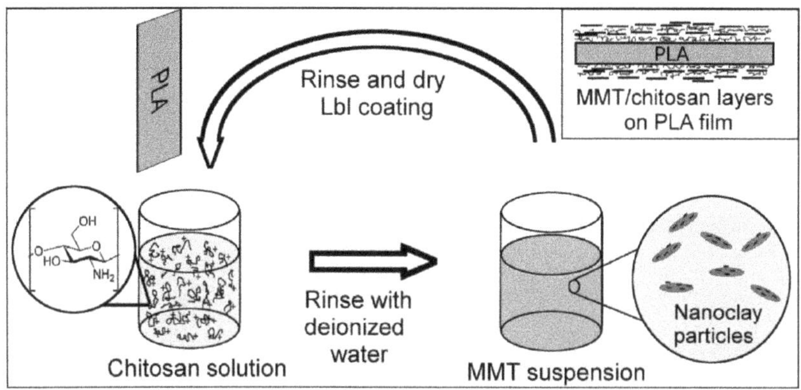

Figure 10.7 Schematic of layer by layer deposition process for the generation of multilayer MMT/chitosan structure on PLA films. Reproduced from Reference 86 with permission from American Chemical Society.

10.6 Vegetable Oil Based Coatings

The economic as well as the environmental concerns of petroleum based materials leads to the necessity of finding out new generation of polymeric products from the raw materials like non-edible vegetable oils and their fatty acids. Vegetable oils (VO) are low cost, non-toxic, biodegradable and renewable materials available aplenty in nature and, thus, are ideal competitors to fossil derived raw materials [94]. The unique structure of VOs with unsaturation sites, esters, hydroxyls, epoxies and other functional groups enable them to undergo chemical modifications to form polymeric materials suitable for various applications especially as a main content in paints and coating formulations [95-101]. VO such as linseed, rubber seed, soybean, karanja oil, castor, cashew nut shell liquid, are being used for the preparation of alkyds [102], epoxies [103,104], polyols [105,106], polyurethanes [107,108], polyesteramides [109,110] and polyetheramides [111,112]. The polymeric formulations derived from vegetable oils reduce or completely avoid the use of volatile organic solvents (VOCs), thus, resulting in high solid, waterborne coatings with environmentally friendly nature. Recently, Alam *et al.* [94] reviewed in detail the vegetable oil based coating materials [94]. The review contained a brief description of role of VO as corrosion inhibitors and polymeric binders in coatings. Another review from Sharmin *et al.*

[113] described in detail the modification of VO for the fabrication of eco-friendly hyperbranched, solvent free, water borne, high solids and UV curable coatings. Anand *et al.* [114] reported the preparation of polyurethane (PU) coatings using renewable material like sorbitol, diacids and 1,4-butanediol for the polyester polyol part of PU. Each

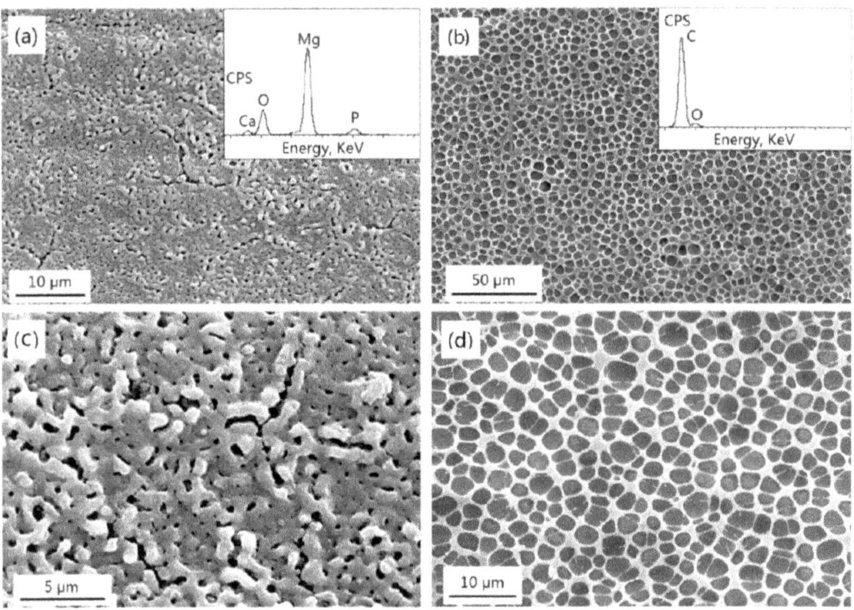

Figure 10.8 SEM morphologies of (a) the MAO coatings and (b) the MAO/PLLA composite coatings, (c and d) the high magnification of panels a and b, respectively. Reproduced from reference 87 with permission from American Chemical Society.

component was selected based on their unique properties. The poly-hydroxy structure provided high thermal, mechanical and solvent resistance properties to sorbitol. Renewable as well as non-renewable acids were selected based on their chain length and molecular weight. PU anti-corrosion coatings were prepared by mixing polyester polyol and 4,4′-methylenebis (phenylisocyanate) (MDI) in cyclohexanone keeping a 1.1: 1 ratio for NCO/OH and subsequently brushcoated on the mild steel substrate. Coatings prepared from long chain diacids like sebacic acid exhibited enhanced physical and chemical resistance to corrosion and better gloss and pencil hardness. Chaudhari *et al.* [115] fabricated an environment friendly polyurethane coating using neem oil polyetheramide. Initially, neem oil was allowed to react with diethanol amine to form fatty amide. The obtained fatty amide was modified to form polyetheramide by reacting it with bisphenol-A. Polyurethane coatings were prepared from the polyetheramide by treating it with methylene diphenyl diisocyanate. Nano-TiO$_2$ modified using silane coupling agents were also added from 0 to 4%. PU coatings with increased chemical resistance and thermal stability were obtained by the incorporation of nano-TiO$_2$ which increased the gloss and pencil hardness significantly. Neem seed oil is mainly used for pharmaceutical applications. Therefore, the current development in the use of need seed oil for the synthesis of high performing PU coatings confirms the potential utilization of renewable resources for the coatings suitable for industrial applications.

Petrovic and Javin [116] used epoxidized soybean oil with 4,4 -diphenylmethane diisocyanate (DMDI) polymer to obtain soybean oil based polyols which produced polyurethane coatings with strong adhesion and hardness. The authors introduced hydroxyl functional groups on the double bonds of the fatty acid and reported a hydroxyl number higher than 250 mg KOH/g. However, the polyols derived from vegetable oil had certain drawbacks in the coating applications because of their low glass transition temperature and low modulus. These resulted from the presence of pendant dangling chains as well as high molecular mobility of fatty acids. Therefore, it is required to develop means to reduce the content of pendant chains derived from fatty acids in the production of PU coatings. Kong *et al.* [117,118] attempted to prepare polyols having low molecular weight and high hydroxyl number with low viscosity in order to prepare strong PU polymer suitable for coating and paint applications. The authors prepared eco-friendly poly(ether ester) polyols using 1,2-propanediol/1,3-propanediol and canola oil. The compounds were subjected to epoxidation and hydroxylation (esterification) reactions which led to the ring opening of epoxide groups and transesterification of glycerides with diols, thus, resulting in a more hydroxyl functionalized polyols. The PU generated from such polyols contained less pendant chains and formed a highly crosslinked structure with a high T$_g$ and modulus. The authors also synthesized polyols (LiprolTM) with high hydroxyl numbers from canola (refined), sunflower (refined), camelina, linola flax and NuLin flax for the production of high solid PU coatings [119]. The structure and properties of various polyols depended on the degree of unsaturation, fatty acid content of the starting oils and extent of oligomerization. The PU coatings were prepared by reacting polyols with petrochemical derived diisocyanate (polymeric aromatic diphenylmethylene diisocyanate (pMDI)) and other additives. PU coatings with high T_g, high contact angle, better corrosion resistance and high cross-linking density were obtained which have strong potential to replace the commercial PU materials derived from petroleum byproducts.

Smart polyurethane coatings encapsulated with corrosion inhibitor were prepared by Marathe *et al.* [120] from neem oil. The authors incorporated quinoline as a corrosion inhibitor in the PU coating generated from neem oil acetylated polyester polyol (NAPP). The smart coatings containing 3% microcapsules exhibited a better corrosion protection in both aqueous and acid media. This method of preparing smart coatings derived from bio-based polyols encapsulated with corrosion inhibitors is envisaged to provide a self-healing character to the polyurethane coatings and can be applied for upgrading other vegetable oil based coatings.

Rajput *et al.* [121] prepared polyols by substituting almost all components of petroleum feed stock with renewable sources. The authors used dimer fatty acid and palmitic acid as the renewable source for PU production. Initially, palmitamide was prepared by amidation reaction of palmitic acid. It was then reacted with dimer fatty acid to form a polyesteramide resin (PEPAD). PU formulations were subsequently prepared by reacting a polymeric hexamethylene diisocyanate (Desmodur N 75 BA/X) with a solution of 50% PEPAD in xylene at different NCO/OH ratios (0.9, 1.1 and 1.3). Transparent PU coatings with high thermal stability were obtained for the PU sample containing 1.1:1 NCO/OH content. On increasing the mole ratio, enhanced properties such as corrosion resistance, thermal stability and transparency was observed due to the formation of urea linkages.

In addition to the PU coatings, VO are also used for the development of alkyd resins which are used widely for the protective coatings since 1940 [122]. These resins are VO based polyesters and possess high thermal stability and good color retention. By changing the fatty acid chain length and suitable chemical modification, the range of applications of alkyd coatings can be enhanced. It has been reported that the physical drying and acid/alkali resistance of alkyds can be improved by alkylation reactions [123-125]. Linseed oil and soy oil are the commonly used VOs for preparing alkyd resins. These oils have a fast drying process and the unsaturation sites form polymeric networks by reaction with oxygen which enhances the properties suitable for protective coatings. The literature studies have confirmed the protection provided by the linseed oil based alkyds to be better than soy oil. Araujo *et al.* [125] confirmed this behavior by conducting a comparison study on the anti-corrosion property of alkyd resin modified with linseed oil and soy oil in accelerated conditions and in marine or industrial atmospheric exposure. Linseed oil based coating provided good adhesion and corrosion resistance in the potentiodynamic and EIS analysis.

Poly(ester amide), consisting of amide (–NHCOR) and ester (–COOR) groups on the polymer backbone, is another class of systems derived from vegetable oil based renewable resources. Mesua ferrea seed oil, linseed oil, coconut oil, Pongamia glabra oil, and Ricinus communis are the commonly used VOs for the synthesis of linear poly(ester amide). Hyperbranched poly(ester amide) (HBPEA) derived from VO is a new class of materials having the properties of barrier coatings because of their unique features of high functionality, low melt and solution viscosity and the three dimensional non-entangled structure [126,127]. An ABx (x ≥ 2) type monomers were used for the synthesize of HBPEA polymer resins. Pramanik *et al.* [128] synthesized HBPEA for the first time using N,N -bis(2-hydroxy ethyl) castor oil fatty amide, isophthalic acid, phthalic anhydride and maleic anhydride, as A2 monomers, whereas a diethanol amine was used as B3 monomer. The degree of branching (DB) enhanced the thermal stability and flow behavior of the HBPEA polymers. The coatings exhibited high adhesion strength, mechanical properties, scratch hardness, impact strength and abrasion resistance desirable for the polymeric applications as surface coatings.

Palm oil is another VO used for the preparation of alkyd films. This non-drying oil, however, does not produce air oxidized coherent films due to its small iodine value. Ataei *et al.* [129] reported the palm oleic acid based barrier coatings with fast physical drying and high water and salt resistance. Palm oleic acid based alkyd were prepared by combining oleic acid and glycerol with phthalic anhydride (PA). The coatings were prepared by free-radical polymerization of alkyd with methylmethacrylate (MMA). The coatings exhibited a decreased drying time and increased alkali resistance with the increment of MMA content. Furthermore, the film hardness and adhesion strength of the coating improved significantly by the alkyd incorporation in the coating formulation.

In summary, it can be concluded that the vegetable oil based coatings have a good potential to be used as eco-friendly resins in surface coatings and can gradually substitute the petroleum based binders. Further developments will enhance the opportunities for making high performance coatings suitable for metal substrates exposed to aggressive environmental conditions at industrial areas.

References

1. Gonzalez-Garcia, Y., Gonzalez, S., and Souto, R. M. (2007) Electrochemical and structural properties of a polyurethane coating on steel substrates for corrosion protection. *Corrosion Science*, **49**, 3514-3526.
2. Leidheiser, H. (1982) Corrosion of painted metals - A review. *Corrosion*, **38**, 374-383.
3. Walter, G. W. (1986) A critical review of the protection of metals by paints. *Corrosion Science*, **26**, 27-38.
4. Chaudhari, A., Kuwar, A., Mahulikar, P., Hundiwale, D., Kulkarni, R., and Gite, V. (2014) Development of anticorrosive two pack polyurethane coatings based on modified fatty amide of Azadirachta indica Juss oil cured at room temperature – a sustainable resource. *RSC Advances*, **4**, 17866-17872.
5. Rajput, S. D., Mahulikar, P. P., and Gite, V. V. (2014) Biobased dimer fatty acid containing two pack polyurethane for wood finished coatings. *Progress in Organic Coatings*, **77**, 38-46.
6. Meshram, P. D., Puri, R. G., Patil, A. L., and Gite, V. V. (2013) Synthesis and characterization of modified cottonseed oil based polyesteramide for coating applications. *Progress in Organic Coatings*, **76**, 1144-1150.
7. Gandini, A., and Belgacem, M. N. (2002) Recent contributions to the preparation of polymers derived from renewable resources. *Journal of Polymers and Environment*, **10**, 105-114.
8. Derksen, J. T. P., Cuperus, F. P., and Kolster, P. (1996) Renewable resources in coatings technology: a review. *Progress in Organic Coatings*, **27**, 45-53.
9. Mahanta, A. K., Mittal, V., Singh, N., Dash, D., Malik, S., Kumar, M., and Maiti, P. (2015) Polyurethane-grafted chitosan as new biomaterials for controlled drug delivery. *Macromolecules*, **48**(8), 2654-2666.
10. Umoren, S. A., and Ekanem, U. F. (2010) Inhibition of mild steel corrosion in H_2SO_4 using exudate gum from *pachylobus edulis* and synergistic potassium halide additives. *Chemical Engineering Communications*, **197**, 1339-1356.
11. Umoren, S. A., and Eduok, U. M. (2016) Application of carbohydrate polymers as corrosion inhibitors for metal substrates in different media: A review. *Carbohydrate Polymers*, **140**, 314-341.
12. Bayol, E., Gurten, A. A., Dursun, M., and Kayakirilmaz, K. (2008) Adsorption behavior and inhibition corrosion effect of sodium carbox-

ymethyl cellulose on mild steel in acidic medium. *Acta Physico-Chimica Sinica*, **24**(12), 2236–2242.

13. Solomon, M. M., Umoren, S. A., Udosoro, I. I., and Udoh, A. P. (2010) Inhibitive and adsorption behaviour of carboxymethyl cellulose on mild steel corrosion in sulphuric acid solution. *Corrosion Science*, **52**(4), 1317-1325.

14. Rajeswari, V., Kesavan, D., Gopiraman, M., Viswanathamurthi, P. (2013) Physicochemical studies of glucose, gellan gum, and hydroxypropyl cellulose – Inhibition of cast iron corrosion. *Carbohydrate Polymers*, **95**(1), 288-294.

15. Arukalam, I. O., Madu, I. O., Ijomah, N. T., Ewulonu, C. M., and Onyeagore, G. N. (2014) Acid corrosion inhibition and adsorption behaviour of ethyl hydroxyethyl cellulose on mild steel corrosion. *Journal of Materials*, DOI:10.1155/2014/101709.

16. Arukalam, I. O., Madufo, I. C., Ogbobe, O., and Oguzie, E. E. (2014) Adsorption and inhibitive properties of hydroxypropyl Methylcellulose on the acid corrosion of mild steel. *International Journal of Applied Sciences and Engineering Research*, **3**(1), 241-256.

17. Arukalam, I. O., Madufor, I. C., Ogbobe, O., and Oguzie, E. E. (2015) Inhibition of mild steel corrosion in sulfuric acid medium by hydroxyethyl cellulose. *Chemical Engineering Communications*, **202**, 112-122.

18. Mobin, M., and Rizvi, M. (2017) Adsorption and corrosion inhibition behavior of hydroxyethyl cellulose and synergistic surfactants additives for carbon steel in 1 M HCl. *Carbohydrate Polymers*, **156**, 202-214.

19. Arukalam, I. O., Madufor, I. C., Ogbobe, O., and Oguzie, E. E. (2013) Adsorption and inhibitive properties of hydroxypropyl methylcellulose on the acid corrosion of mild steel. *International Journal of Applied Sciences and Engineering Research*, **2**(6), 613-629.

20. Rajeswari, V., Kesavan, D., Gopiraman, M., and Viswanathamurthi P. (2013) Physicochemical studies of glucose, gellan gum, and hydroxypropyl cellulose - Inhibition of cast iron corrosion. *Carbohydrate Polymers*, **95**, 288-294.

21. Umoren, S. A., Solomon, M. M., Udosoro, I. I., and Udoh, A. P. (2010) Synergistic and antagonistic effects between halide ions and carboxymethyl cellulose for the corrosion inhibition of mild steel in sulphuric acid solution. *Cellulose*, **17**(3), 635-648.

22. Manimaran, N., Rajendran, S., Manivannan, M., Thangakani, J. A., and Suriya Prabha A. (2013) Corrosion Inhibition by carboxymethyl cellulose. *European Chemical Bulletin*, **2**(7), 494-498.

23. Rajendran, S., Sridevi, S. P., Anthony, N., Amalraj, A. J., and Sundearavadivelu, M. (2005) Corrosion behavior of carbon steel in polyvinyl alcohol. *Anti-Corrosion Methods and Materials*, **52**(2), 102-107.

24. Li, M. M., Xu, Q. J., Han, J., Yun, H., and Min, Y. L. (2015) Inhibition

action and adsorption behavior of green inhibitor sodium carboxymethyl cellulose on copper. *International Journal of Electrochemical Science*, **10**, 9028-9041.

25. Tian, H., Li, W., and Hou, B. (2011) Novel application of a hormone biosynthetic inhibitor for the corrosion resistance enhancement of copper in synthetic seawater. *Corrosion Science*, **53**, 3435-3445.

26. Yang, Q., Fukuzumi, H., Saito, T., Isogai, A., and Zhang, L. (2011) Transparent cellulose films with high gas barrier properties fabricated from aqueous alkali/urea solutions. *Biomacromolecules*, **12**(7), 2766-2771.

27. Qi, Z. D., Saito, T., Fan, Y., and Isogai, A. (2012) Multifunctional coating films by layer-by-layer deposition of cellulose and chitin nanofibrils. *Biomacromolecules,* **13**(2), 553-558.

28. Wefers, K., Nitowski, G. A., and Wieserman, L. F. (1992) Phosphonic/Phosphinic Acid Bonded to Aluminum Hydroxide Layer, US Patent 5132181.

29. Arai, T., Shin, R. B., Yamamoto, T. (2008) Agents for the Surface Treatment of Zinc or Zinc Alloy Products, US Patent 20080113102.

30. Walters, D. N., Schneider, J. R. (2008) Coating Compositions Exhibiting Corrosion Resistance Properties and Related Coated Substrates, US Patent 20080090069.

31. Tamborim, S. M., Dias, S. L. P., Silva, S. N., Dick, L.F.P., and Azambuja, D.S. (2011) Preparation and electrochemical characterization of amoxicillin-doped cellulose acetate films for AA2024-T3 aluminum alloy coatings. *Corrosion Science*, **53**, 1571-1580.

32. Abdallah, M. (2004) Antibacterial drugs as corrosion inhibitors for corrosion of aluminium in hydrochloric solution. *Corrosion Science*, **46**, 1981-1996.

33. Yabuki, A., Kawashim, A., and. Fathon, I. W. (2014) Self-healing polymer coatings with cellulose nanofibers served as pathways for the release of a corrosion inhibitor. *Corrosion Science*, **85**, 141-146.

34. Yabuki, A., and Nishisaka, T. (2011) Self-healing capability of porous polymer film with corrosion inhibitor inserted for corrosion protection. *Corrosion Science*, **53**, 4118-4123.

35. Yabuki, A., Shiraiwa, T., Fathon, I. W. (2016) pH-controlled self-healing polymer coatings with cellulose nanofibers providing an effective release of corrosion inhibitor. *Corrosion Science*, **103**, 117-123.

36. Hirano, S., Inui, H., Kosaki, H., Uno, Y., and Toda, T. (1994) Chitin and chitosan: ecologically bioactive polymers. In: *Biotechnology and Bioactive Polymers*, Gebelein, C. G., and Carraher, Jr., C. E. (eds.), Plenum Press, New York, pp. 43-54.

37. Nishiyama, Y., Langan, P., and Chjanzy, H. (2002) Crystal structure and hydrogen-bonding system in cellulose Iß from synchrotron x-ray and neutron fiber diffraction. *Journal of the American Chemical Society*, **124**, 9074-9082.

38. Sandford, P. A., and Steinners, A. (1991). Biomedical applications of high-purity chitosan. In: *Water-soluble Polymers*, Shalaby, S. W., Cormick, C. L., and Butles, G. B. (eds.), ACS Symposium Series 467, Washington, USA, p. 430.

39. Eduok, U. M., and Khaled, M. M. (2014) Retraction Note to: Corrosion protection of steel sheets by chitosan from shrimp shells at acid pH. *Cellulose*, **21**, 3139-3143.

40. Vathsala, K., Venkatesha, T. V., Praveen, B. M., and Nayana, K. O. (2010) Electrochemical generation of Zn-chitosan composite coating on mild steel and its corrosion studies. *Engineering*, **2**, 580-584.

41. El-Haddad, M. N. (2013) Chitosan as a green inhibitor for copper corrosion in acidic medium. *International Journal of Biological Macromolecules*, **55**, 142-149.

42. Solmaz, R., Kardas, G., Yazıcı, B., and Erbil, M. (2008) Adsorption and corrosion inhibitive properties of 2-amino-5-mercapto-1,3,4-thiadiazole on mild steel in hydrochloric acid media. *Colloids and Surfaces A: Physicochemical and Engineering Aspects*, **312**, 7-17.

43. Fekry, A. M., and Mohamed, R. R. (2010) Acetyl thiourea chitosan as an eco-friendly inhibitor for mild steel in sulphuric acid medium. *Electrochimica Acta*, **55**, 1933-1939.

44. Cheng, S., Chen, S., Liu, T., Chang, X., and Yin, Y. (2007) Carboxymethylchitosan + Cu^{2+} mixture as an inhibitor used for mild steel in 1 M HCl. *Electrochima Acta*, **52**, 5932-5938.

45. Mohamed, R. R., and Fekry, A. M. (2011) Antimicrobial and anticorrosive activity of adsorbents based on chitosan Schiff's base. *International Journal of Electrochemical Science,* **6,** 2488-2508.

46. Goissis, G., Junior, E. M., Marcantonio, R. A. C., Lai, R. C. C., Cancian, D. C. J., and DeCaevallho, W. M. (1999) Biocompatibility studies of anionic collage membranes with different degree of glutaraldehyde cross-linking. *Biomaterials*, **20**, 27-34.

47. Huang, R. Y. M., Pal, R., and Moon, G. Y. (1999) Crosslinked chitosan composite membrane for the pervaporation dehydration of alcohol mixtures and enhancement of structural stability of chitosan/polysulfone composite membranes. *Journal of Membrane Science*, **160**, 17-30.

48. Wei, Y. C., Hudson, S. M., Mayer, J. M., and Kaplan, D. L. (1992) The crosslinking of chitosan fibers. *Journal of Polymer Science, Part A: Polymer Chemistry*, **30**, 2187-2193.

49. Schmidt, C. E., and Baier, J. M. (2000) Acellular vascular tissues: natural biomaterials for tissue repair and tissue engineering. *Biomaterials*, **21**, 2215-2231.

50. Luckachan, G. E., and Mittal, V. (2015) Anti-corrosion behavior of layer by layer coatings of cross-linked chitosan and poly(vinyl butyral) on carbon steel. *Cellulose,* **22**, 3275-3290.

51. Luckachan, G. E., and Mittal, V. Stable corrosion-resistant chitosan

coatings: effect of crosslinking with silica, poly(vinyl butyral) over coating and graphene on the coating performance, in preparation.

52. Liu, Y., Zou, C., Yan, X., Xiao, R., Wang, T., and Li, M. (2015) β-Cyclodextrin modified natural chitosan as a green inhibitor for carbon steel in acid solutions. *Industrial & Engineering Chemistry Research,* **54**(21), 5664-5672.

53. Hoagland, P. D., and Parris, N. (1996) Chitosan/pectin laminated films. *Journal of Agricultural and Food Chemistry,* **44**, 1915-1919.

54. Szymanska, E., and Winnicka, K. (2015) Stability of chitosan – a challenge for pharmaceutical and biomedical applications. *Marine Drugs*, **13**, 1819-1846.

55. Tian, D., Dubois, P. H., Grandfile, C. H., Jermome, P., Viville, P., Lazzaroni, R., Bredas, J. L., and Leprince, P. (1997) A novel biodegradable and biocompatible ceramer prepared by the sol–gel process. *Chemistry of Materials*, **9**, 871-874.

56. Sugama, T., and Milian-Jimenez, S. (1999) Dextrine-modified chitosan marine polymer coatings. *Journal of Material Science*, **34**, 2003-2014.

57. Kumar, G., and Buchheit, R. G. (2006) Development and characterization of corrosion resistant coatings using the natural biopolymer chitosan. *ECS Transactions*, **1**(9), 101-117.

58. Sugama, T., and Cook, M. (2000) Poly(itaconic acid)-modified chitosan coatings for mitigating corrosion of aluminum substrates. *Progress in Organic Coatings*, **38**, 79-87.

59. Chui, V. W. D., Mok, K. W., Ng, C. Y., Luong, B. P., and Ma, K. K. (1996) Removal and recovery of copper(II), chromium(III), and nickel(II) from solutions using crude shrimp chitin packed in small columns. *Environment International*, **22**, 463-468.

60. Schlick, S. (1986) Binding sites of copper$_{2+}$ in chitin and chitosan. An electron spin resonance study. *Macromolecules*, **19**, 192-195.

61. Domard, A. (1987) pH and c.d. measurements on a fully deacetylated chitosan: application to Cu[II]—polymer interactions. *International Journal of Biological Macromolecules*, **9**, 98-104.

62. Nieto, J. M., Covas, C. P., and Del Bosque, J. (1992) Preparation and characterization of a chitosan-Fe(III) complex. *Carbohydrate Polymers*, **18**, 221-224.

63. Lundvall, O., Gulppi, M., Paez, M. A., Gonzalez, E., Zagal, J. H., Pavez, J., and Thompson, G. E. (2007) Copper modified chitosan for protection of AA-2024. *Surface & Coatings Technology*, **201**, 5973-5978.

64. Zheludkevich, M. L., Tedim, J., Freire, C. S. R., Fernandes, S. C. M., Kallip, S., Lisenkov, A., and. Ferreira, M. G. S. (2011) Self-healing protective coatings with "green" chitosan based pre-layer reservoir of corrosion inhibitor. *Journal of Material Chemistry*, **21**, 4805-4812.

65. Carneiro, J., Tedim, J., Fernandes, S. C. M., Freire, C. S.R., Gandini, A., Ferreira, M. G. S., and Zheludkevich, M. L. (2013) Functionalized chi-

tosan-based coatings for active corrosion protection. *Surface & Coatings Technology*, **226**, 51-59.

66. Wang, Y., Dong, C., Zhang, D., Ren, P., Li, L., and Xiao-gang, L. (2015) Preparation and characterization of a chitosan-based low-pH-sensitive intelligent corrosion inhibitor. *International Journal of Minerals, Metallurgy and Materials*, **22**, 998-1004.

67. Carneiro, J., Tedim, J., Fernandes, S. C. M., Freire, C. S. R., Gandini, A., Ferreira, M. G. S., and Zheludkevich, M. L. (2013) Chitosan as a smart coating for controlled release of corrosion inhibitor 2-mercapto-benzothiazole. *ECS Electrochemistry Letters*, **2**(6), C19-C22.

68. Carneiro, J., Tedim, J., Fernandes, S. C. M., Freire, C. S. R., Silvestre, A. J. D., Gandini, A., Ferreira, M. G. S., and Zheludkevich, M. L. (2012) Chitosan-based self-healing protective coatings doped with cerium nitrate for corrosion protection of aluminum alloy 2024. *Progress in Organic Coatings*, **75**, 8-13.

69. Hassannejad, H., and Nouri, A. (2016) Synthesis and evaluation of self-healing cerium-doped chitosan nanocomposite coatings on AA5083-H321. *International Journal of Electrochemical Science*, **11**, 2106-2118.

70. Franchin, C., Muraglia, M., Corbo, F., Florio, M. A., Di Mola, A., Rosato, A., Matucci, R., Nesi, M., Bambeke, F. V., and Vitali, C. (2009) Synthesis and biological evaluation of 2-mercapto-1,3-benzothiazole derivatives with potential antimicrobial activity. *Archiv der Pharmazie*, **342**, 605-613.

71. Bao, Q., Zhang, D., and Wan, Y. (2011) 2-Mercaptobenzothiazole doped chitosan/11-alkanethiolate acid composite coating: Dual function for copper protection. *Applied Surface Science*, **257**, 10529-10534.

72. Fayyad, E. M., Sadasivuni, K. K., Ponnamma, D., and Al-Maadeed, M. A. A. (2016) Oleic acid-grafted chitosan/graphene oxide composite coating for corrosion protection of carbon steel. *Carbohydrate Polymers*, **151**, 871-878.

73. Sugama, T., and DuVall, J. E. (1996) Polyorganosiloxane-grafted potato starch coatings for protecting aluminum from corrosion. *Thin Solid Films*, **289**, 39-48.

74. Rosliza, R., and Wan Nik, W. B. (2010) Improvement of corrosion resistance of AA6061 alloy by tapioca starch in seawater. *Current Applied Physics*, **10**, 221-229.

75. Gao, B., Zhang, X., and Sheng, Y. (2008) Studies on preparing and corrosion inhibition behaviour of quaternized polyethyleneimine for low carbon steel in sulfuric acid. *Materials Chemistry and Physics*, **108**, 375-381.

76. Al-Juhni, A. A., and Newby, B. Z. (2006) Incorporation of benzoic acid and sodium benzoate into silicone coatings and subsequent leaching of the compound from the incorporated coatings. *Progress in Or-*

ganic Coatings, **56**, 135-145.

77. Abd El Haleem, S. M. (1986) Anodic behaviour and pitting corrosion of plain carbon steel in NaOH solutions containing Cl⁻ ions. *Surface and Coatings Technology*, **27**, 167-173.

78. Bello, M., Ochoa, N., Balsamo, V., López-Carrasquero, F., Coll, S., Monsalve, A., and Gonzalez, G. (2010) Modified cassava starches as corrosion inhibitors of carbon steel: An electrochemical and morphological approach. *Carbohydrate Polymers*, **82**, 561-568.

79. Ochoa, N., Bello, M., Sancristóbal, J., Balsamo, V., Albornoz, A., and Brito, J. L. (2013) Modified cassava starches as potential corrosion inhibitors for sustainable development. *Materials Research*, **16**(6), 1209-1219.

80. Carrasco, F., Pages, P., Gamez-Perez, J., Santana, O. O., and Maspoch, M. L. (2010) Kinetics of the thermal decomposition of processed poly(lactic acid). *Polymer Degradation and Stability*, **95**, 2508-2514.

81. Lasprilla, A. J. R., Martinez, G. A. R., Lunelli, B. H., Jardini, A. L., and Filho, R. M. (2012) Poly-lactic acid synthesis for application in biomedical devices - A review. *Biotechnology Advances*, **30**, 321-328.

82. Nampoothiri, K. M., Nair, N. R., and John, R. P. (2010) An overview of the recent developments in polylactide (PLA) research. *Bioresource Technology*, **101**, 8493–8501.

83. Gupta, B., Revagadea, N., and Hilborn, J. (2007) Poly(lactic acid) fiber: An overview. *Progress in Polymer Science*, **32**, 455-482.

84. Lim, L. T., Auras, R., and Rubino, M. (2008) Processing technologies for poly(lactic acid). *Progress in Polymer Science*, **33**, 820–852.

85. Carrasco, F., Pagesb, P., Gamez-Perez, J., Santana, O. O., and Maspoch, M. L. (2010) Processing of poly(lactic acid): Characterization of chemical structure, thermal stability and mechanical properties. *Polymer Degradation and Stability*, **95**, 116-125.

86. Svagan, A. J., Akesson, A., Cardenas, M., Bulut, S., Knudsen, J. C., Risbo, J., and Plackett, D. (2012) Transparent films based on PLA and montmorillonite with tunable oxygen barrier properties. *Biomacromolecules*, **13**(2), 397-405.

87. Zeng, R. C., Cui, L. Y., Jiang, K., Liu, R., Zhao, B. D., and Zheng, Y. F. (2016) In vitro corrosion and cytocompatibility of a microarc oxidation coating and poly(L-lactic acid) composite coating on Mg-1Li-1Ca alloy for orthopedic implants. *ACS Applied Materials & Interfaces*, **8**(15), 10014-10028.

88. Ocak, Y., Aysun, S., Funda, T., and Hasan B. (2009) Protection of marble surfaces by using biodegradable polymers as coating agent. *Progress in Organic Coatings*, **66**, 213-220.

89. Ocak, Y., Sofuoglu, A., Tihminlioglu, F., and Boke, H. (2015) Sustainable bio-nanocomposite coatings for the protection of marble surfaces. *Journal of Cultural Heritage,* **16**(3), 299-306.

90. Frediani, M., Rosi, L., Camaiti, M., Berti, D., Mariotti, A., Comucci, A.,

Vannucci, C., and Malesci, I. (2010) Polylactide/perfluoropolyether block copolymers: Potential candidates for protective and surface modifiers. *Macromolecular Chemistry and Physics*, **211**, 988-995.

91. Giuntoli, G., Rosi, L., Frediani, M., Sacchi, B., and Frediani, P. (2012) Fluoro-functionalized PLA polymers as potential water-repellent coating materials for protection of stone. *Journal of Applied Polymer Science*, **125**, 3125-3133.

92. Giuntoli, G., Frediani, M., Pedna, A., Rosi, L., and Frediani, P. (2012) New perspectives for the application of PLA in cultural heritage. In: *Polylactic Acid: Synthesis, Properties and Applications*, Nova Science Publishers, USA, pp.161-189.

93. Pedna, A., Pinho, L., Frediani, P., Mosquera, M. J. (2016) Obtaining SiO_2–fluorinated PLA bionanocomposites with application as reversible and highly-hydrophobic coatings of buildings. *Progress in Organic Coatings*, **90**, 91-100.

94. Alam, M., Akram, D., Sharmin, E., Zafar, F., and Ahmad, S. (2014) Vegetable oil based eco-friendly coating materials: A review article. *Arabian Journal of Chemistry*, **7**, 469-479.

95. Dutton, H. J., and Scholfield, C. R. (1963) Recent developments in the glyceride structure of vegetable oils. *Progress in the Chemistry of Fats and other Lipids*, **6**, 313-339.

96. Baumann, H., Buhler, M., Fochem, H., Hirsinger, F., Zoebelein, H., and Falbe, J. (1988) Natural fats and oils - Renewable raw materials for the chemical industry. *Angewandte Chemie International Edition*, **27**, 41-62.

97. Wisniak, J. (1977) Jojoba oil and derivatives. *Progress in the Chemistry of Fats and other Lipids*, **15**, 167-218.

98. Schuchardt, U., Sercheli, R., and Vargas, R. M. (1998) Transesterification of vegetable oils: A review. *Journal of the Brazilian Chemical Society*, **9**, 199-210.

99. Lu, Y., and Larock, R. C. (2009) Novel polymeric materials from vegetable oils and vinyl monomers: Preparation, properties, and applications. *ChemSusChem*, **2**, 136-147.

100. Xia, Y., and Larock, R. C. (2010) Vegetable oil-based polymeric materials: synthesis, properties, and applications. *Green Chemistry*, **12**, 1893-1909.

101. Salimon, J., Salih, N., and Yousif, E. (2012) Industrial development and applications of plant oils and their biobased oleochemicals. *Arabian Journal of Chemistry*, **5**, 135-145.

102. Odetoye, E., Ogunniyi, D. S, and Olatunji, G. A. (2010) Preparation and evaluation of Jatropha curcas Linneaus seed oil alkyd resins. *Industrial Crops and Products*, **32**(3), 225-230.

103. Shikha, D., Kamani, P. K., and Shukla, M. C. (2003) Studies on synthesis of water-borne epoxy ester based on RBO fatty acids. *Progress in Organic Coatings*, **47**(2), 87-94.

104. Ramasri, M., Srinivasa Rao, G. S., Sampathkumaran, P. S., and Shirsalkar, M. M. (1990) Water-soluble epoxy binders modified with boron ester for cathodic electrodeposition. *Progress in Organic Coatings*, **18**(1), 103-115.
105. Argyropoulos, J., Popa, P., Spilman, G., Bhattacharjee, D., and Koonce, W. (2009) Seed oil based polyester polyols for coatings. *Journal of Coating Technology and Research*, **6**(4), 501-508.
106. Laxmikanth Rao, J., Balakrishna, R. S., and Shirsalkar, M. M. (1992) Cathodically electrodepositable novel coating system from castor oil. *Journal of Applied Polymer Science*, **44**(11), 1873-1881.
107. Petrovic, Z. S. (2008) Polyurethanes from vegetable oils. *Polymer Reviews*, **48**(1), 109-155.
108. Lligadas, G., Ronda, J. C., Galia, M., and Cadiz, V. (2010) Plant oils as platform chemicals for polyurethane synthesis: Current state-of-the-art. *Biomacromolecules*, **11**, 2825-2835.
109. Meshram, P. D., Puri, R. G., Patil, A. L, and Gite, V. V. (2013) High performance moisture cured poly (ether–urethane) amide coatings based on renewable resource (cottonseed oil). *Journal of Coating Technology and Research*, **10**, 331-338.
110. Mahapatra, S. S., and Karak, N. (2004) Synthesis and characterization of polyesteramide resins from Nahar seed oil for surface coating applications. *Progress in Organic Coatings*, **51**, 103-108.
111. Alam, M., Sharmin, E., Ashraf, S. M., and Ahmad, S. (2004) Newly developed urethane modified polyetheramide-based anticorrosive coatings from a sustainable resource. *Progress in Organic Coatings*, **50**, 224-230.
112. Gaikwad, M. S., Gite, V. V., Mahulikar, P. P., Hundiwale, D. G., and Yemul, O. S. (2015) Eco-friendly polyurethane coatings from cottonseed and karanja oil. *Progress in Organic Coatings*, **86**, 164-172.
113. Sharmin, E., Zafar, F., Akram, D., Alam, M., and Ahmad, S. (2015) Recent advances in vegetable oils based environment friendly coatings: A review. *Industrial Crops and Products*, **76**, 215-229.
114. Anand, A., Kulkarni, R. D., Patila, C. K., and Gite, V. V. (2016) Utilization of renewable bio-based resources, *viz.* sorbitol, diol, and diacid, in the preparation of two pack PU anticorrosive coatings. *RSC Advances*, **6**, 9843-9850.
115. Chaudhari, A. B., Anand, A., Rajput, S. D, Kulkarni, R. D, and Gite, V. V. (2013) Synthesis, characterization and application of Azadirachta indica juss (neem oil) fatty amides (AIJFA) based polyurethanes coatings: A renewable novel approach. *Progress in Organic Coatings*, **76**, 1779-1785.
116. Petrovic, Z. S., and Javni, I. J. (2002) Process for the Synthesis of Epoxidized Natural Oil-based Isocyanate Prepolymers for Application in Polyurethanes, US Patent 6,399,698.
117. Kong, X. H., Liu, G. G., and Curtis, J. M. (2012) Novel polyurethane

produced from canola oil based poly(ether ester) polyols: Synthesis, characterization and properties. *European Polymer Journal*, **48**, 2097-2106.

118. Anuar, S. T, Zhao, Y. Y., Mugo, S. M., and Curtis, J. M. (2012) Monitoring the epoxidation of canola oil by non-aqueous reversed phase liquid chromatography/mass spectrometry for process optimization and control. *Journal of American Oil Chemists Society*, **89**, 1951-1960.

119. Kong, X., Liu, G., Qi, H., and Curtis, J. M. (2013) Preparation and characterization of high-solid polyurethane coating systems based on vegetable oil derived polyols. *Progress in Organic Coatings*, **76**, 1151-1160.

120. Marathe, R., Tatiya, P., Chaudhari, A., Lee, J., Mahulikar, P., Sohn, D., and Gite, V. (2015) Neem acetylated polyester polyol - Renewable source based smart PU coatings containing quinoline (corrosion inhibitor) encapsulated polyurea microcapsules for enhanced anticorrosive property. *Industrial Crops and Products*, **77**, 239-250.

121. Rajput, S. D., Hundiwale, D. G., Mahulikar, P. P., and Gite, V. V. (2014) Fatty acids based transparent polyurethane films Dilip G and coatings. *Progress in Organic Coatings*, **77**, 1360-1368.

122. Heiskanen, N., Jamsa, S., Paajanen, L., and Koshimies, S. (2010) Synthesis and performance of alkyd–acrylic hybrid binders. *Progress in Organic Coatings*, **67**, 329-338.

123. Akbarinezhad, E., Ebrahimi, M., Kassiriha, S. M., and Khorasani, M. (2009) Synthesis and evaluation of water-reducible acrylic–alkyd resins with high hydrolytic stability. *Progress in Organic Coatings*, **65**, 217–221.

124. Akintayo, C. O., and Adebowale, K. O. (2004) Synthesis and characterization of acrylated Albizia benth medium oil alkyds. *Progress in Organic Coatings*, **50**, 207-212.

125. Araujo, W. S., Margarit, I. C. P., Mattos, O. R., Fragata, F. L., and Lima-Netoe, P. (2010) Corrosion aspects of alkyd paints modified with linseed and soy oils. *Electrochimica Acta*, **55**, 6204-6211.

126. Mezzenga, R., and Manson, J. A. E. (2001) Thermo-mechanical properties of hyperbranched polymer modified epoxies. *Journal of Material Sciences*, **36**, 4883-4891.

127. Pilla, S., Kramschuster, A., Lee, J., Clemons, C., Gong, S., and Turng, LS. (2010) Microcellular processing of polylactide-hyperbranched polyester-nanoclay composites. *Journal of Material Science*, **45**, 2732-2746.

128. Pramanik, S., Konwarh, R., Sagar, K., Konwar, B. K., and Karak, N. (2013) Bio-degradable vegetable oil based hyperbranched poly(ester amide) as an advanced surface coating material. *Progress in Organic Coatings*, **76**, 689-697.

129. Ataei, S., Yahya, R., and Gan, S. N. (2011) Fast drying, high water and

salt resistance coatings from non-drying vegetable oil. *Progress in Organic Coatings*, **72**, 703-708.

Index

A

abrasion resistance, 34, 37-40, 42, 49, 51, 219

actIvity, 48, 108, 131, 147-148, 223, 225

adhesion, 6, 50, 84, 91, 96, 148, 150, 171, 203, 219

adsorbents, 223

B

barrier properties, 47, 56, 65, 93, 199, 209, 222, 226

batteries, 107, 129, 186

biodegradable, 67, 212-213, 215, 224, 226

biofouling, 108

C

capacitors, 105, 114-115, 133

capillary effect, 153

carbonaceous char, 95

catalysis, 88, 129

chain scission, 92-93, 95-96

chemical stability, 33, 91

compatibility, 109-110, 115-116, 134, 172

compatibilizer, 47

contact angle, 210, 212, 214, 218

cracking, 42, 94, 114

■ Index

crystallinity, 57-58, 123

D

degradation temperature, 6, 50-51

dehydration, 92, 96, 223

dispersant, 187, 195

dissociation, 73, 152

drug delivery, 79-80, 220

durability, 35, 68, 125

E

electrical conductivity, 5, 14-15, 43, 50, 105, 107-108, 110, 113, 119-120, 124-125, 140

electrochemical impedance spectroscopy, 101

electrochemical properties, 145-146

electron mobility, 106, 131

encapsulation, 91, 188, 195

energy storage, 107, 130

F

flame retardancy, 2, 10, 12, 70, 102-103

flammable, 95

functionalized graphene, 44, 56, 59, 62, 68, 70, 108, 132

fusion bonded epoxy, 97

G

gas impermeability, 36, 44, 46-47, 213

gene delivery, 79

glass transition temperature, 2, 4-5, 44, 92, 168, 217

graphitic carbon, 181

H

homopolymers, 175

■Index

hydrocarbons, 140
hydrogels, 31
hyperbranched polymers, 184

I

impact strength, 2-3, 57, 219
initiator, 19, 21-24, 28, 30-31, 73, 76-77, 80, 88
interfacial interaction, 38, 61, 208
isomerization, 179

L

layer-by-layer, 110, 132, 163, 190, 200, 222
LCST, 20

M

macroporous, 20, 31, 131
melting temperature, 55
membranes, 52-56, 67-69, 88, 108, 131, 162, 223
monolith, 19-20, 22-23, 25-30, 109

N

nanocrystals, 184

P

PEMFC, 53-55, 108
permeability, 40, 44, 46-47, 51, 53-54, 58, 200, 208-209, 212
plasticizer, 57, 71
polydispersity index, 76
pores, 25
porosity, 19, 25

power density, 53, 55, 107

R

Raman spectra, 203-204
reinforcing fillers, 115
renewable, 154, 213,
    215-220, 227-229
rigidity, 57
roughness, 193, 214

S

selectivity, 56, 87, 153
self-healing polymer
    coatings, 201
sensor materials, 164
shear thinning, 51
silane treatment, 48
sol-gel, 49, 61, 66, 127
specific capacitance, 131
specific surface area, 108
storage modulus, 7, 35,
    38-39, 44, 50, 118, 206-207
surface energy, 19, 214

surface modification, 61, 63,
    66, 172, 181
surfactants, 198, 221

T

tensile strength, 2, 6, 36-39,
    42, 44, 46-47, 49, 51, 55, 57,
    150
thermal conductivity, 5-6,
    15, 38, 44, 64, 106, 113, 124
tissue engineering, 192, 223
toughness, 4, 33, 45, 48-49,
    56-57
tribological properties, 6-7,
    16, 63

U

ultrafiltration, 131

V

vegetable oil, 51, 66, 215, 217-219, 227, 229-230

X

X-ray diffraction, 51

Y

Young's modulus, 35, 39, 44, 113, 151

CPSIA information can be obtained
at www.ICGtesting.com
Printed in the USA
BVHW090611111121
621191BV00002B/174